Published By

Objective Learning Materials
A business of W & G Education Pty Ltd
Unit 1, 7-9 Brough Street,
Springvale, Victoria, 3171
Phone: +61 3 9796 117
Fax: +61 3 9796 1832
www.lat-olm.com.au

Originally published in 1981 by Educational Efficiency Products
ISBN 978-0-9941613-2-1

Printed in Australia By
Objective Learning Materials

MATHOMAT®
INSTRUCTION TEXT BOOK
&
UNITS OF WORK

Published by

All designs and diagrams in this book were constructed with "Mathomat".

1981
Printed in Australia

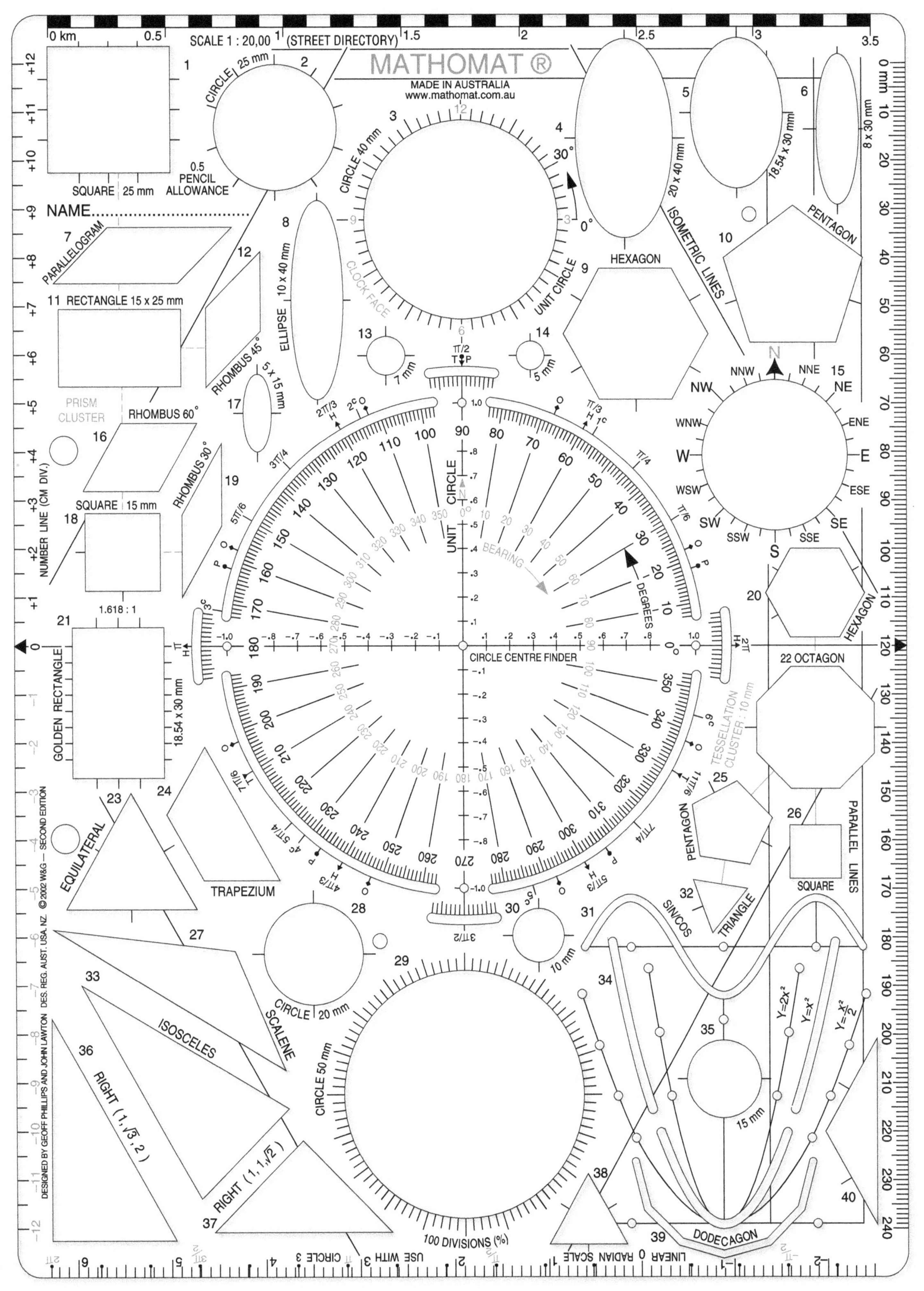

Manufactures and Distributors: Objective Learning Materials, Springvale, Victoria

PREFACE

"Mathomat" Instruction Book, Vol. 2 is a direct follow-on from Vol. 1 and consists of complete unbits of work incorporating the use of the Mathomat template

These units may be used in class or given as home assignments. For example, the unit on ANGLES is given in detail (a teacher can appropriately abbreviate this information) and has been designed expressly for use by students as an assignment to be completed at home and corrected at school. The tests at the end of this unit are given separately in class.

Th unit on SNOWFLAKE GEOMETRY has a sheet (page 3 of this unit) of 12 original snowflake designs. If these designs are examined carefully, it will be seen that they can be enlarged using the appropriate shapes in the Mathomat.

The uni on GEOMETRICAL CREATIVITY has been included to show what can be done with the cut-outs from the Mathomat template. However, each of the diagrams and pattterns can of course be drawn with the Mathomat.

There are times when units such as these can be given to satisfactorily occupy:"

(i) the faster student.

or (ii) a class in the last few weeks of a school year.

I sincerely hope that these units will be found interesting and worthwhile and the same satisfaction obtained using them as I did in creating them. Further units will be continued in "Mathomat" Instruction Book, Vol. 3, to be published at a later date.

Craig Young.

CONTENTS

LINEAR MEASURE

1

1. Use your Mathomat to carefully measure each line in centimetres

eg.

A B C D E F G H

Ans. AB = 3cm CD = cm EF = cm GH = cm

2. Measure these lines in millimetres to the nearest millimetre.

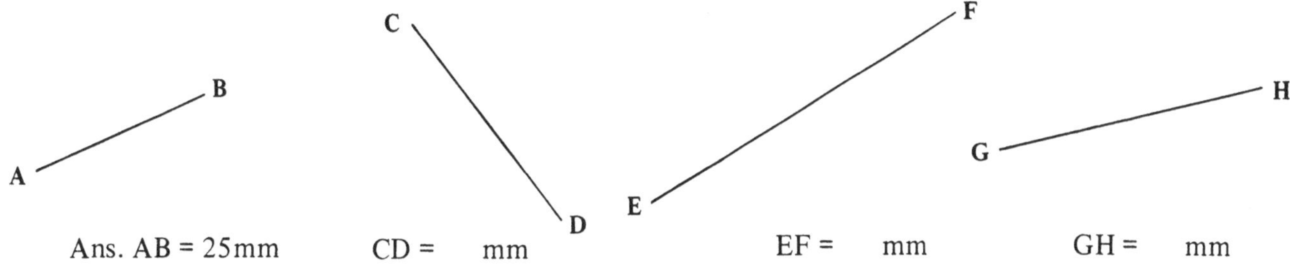

Ans. AB = 25mm CD = mm EF = mm GH = mm

3. Measure these lines in millimetres to the nearest millimetre and then write your answer in centimetres.

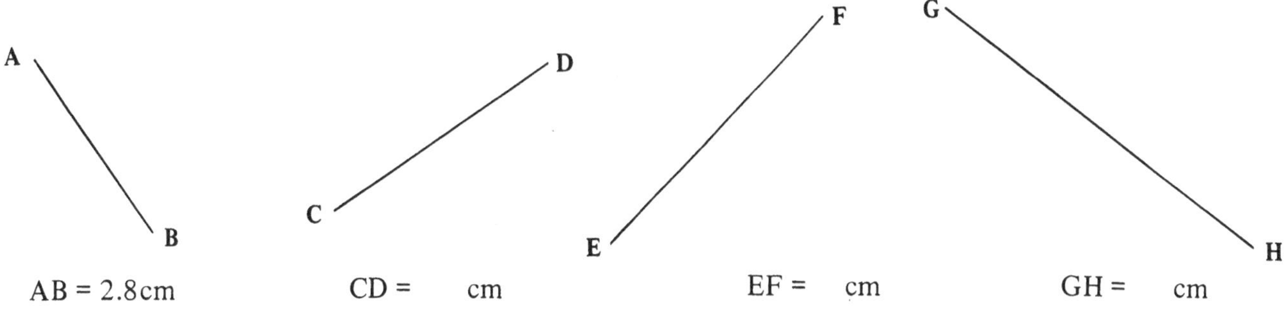

AB = 2.8cm CD = cm EF = cm GH = cm

4. If 1 cm is used to represent 5 kilometres, find the distance from Sandy Bay to the "buried treasure" in kilometres

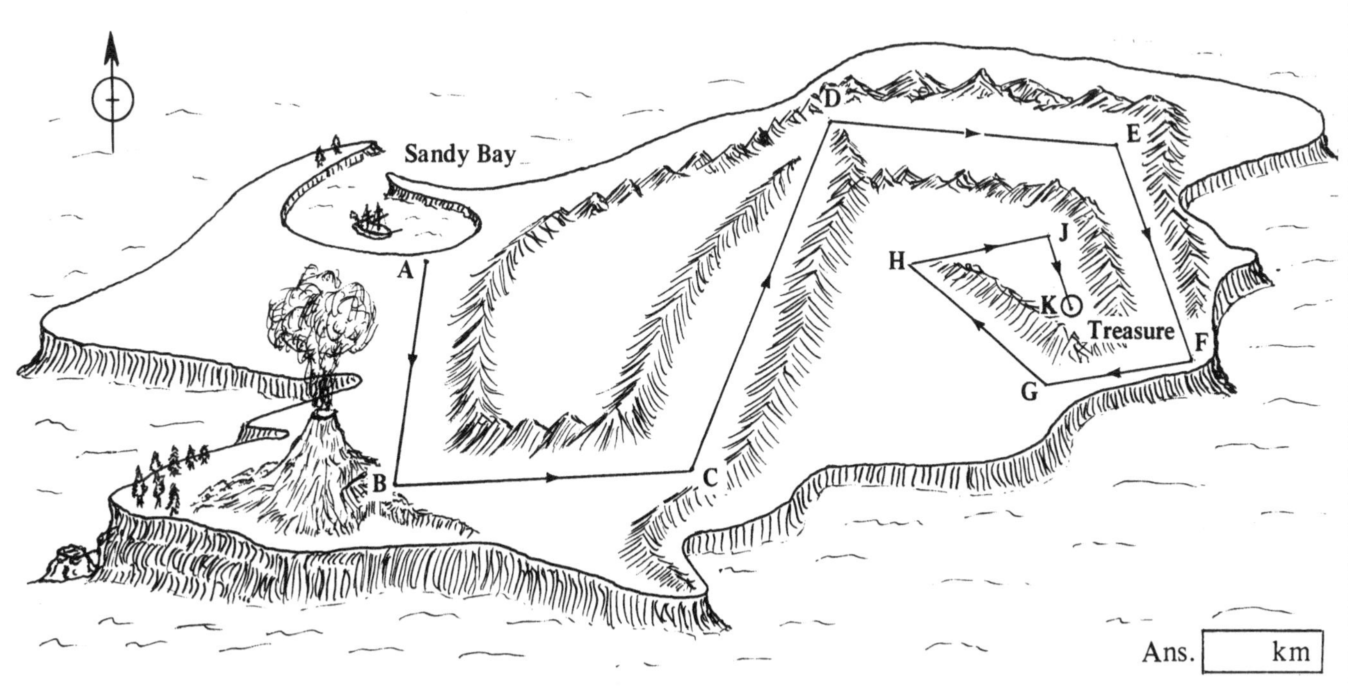

Ans. km

5. If 1 cm is used to represent 1 kilometre, find the distance around this yacht course in kilometres.

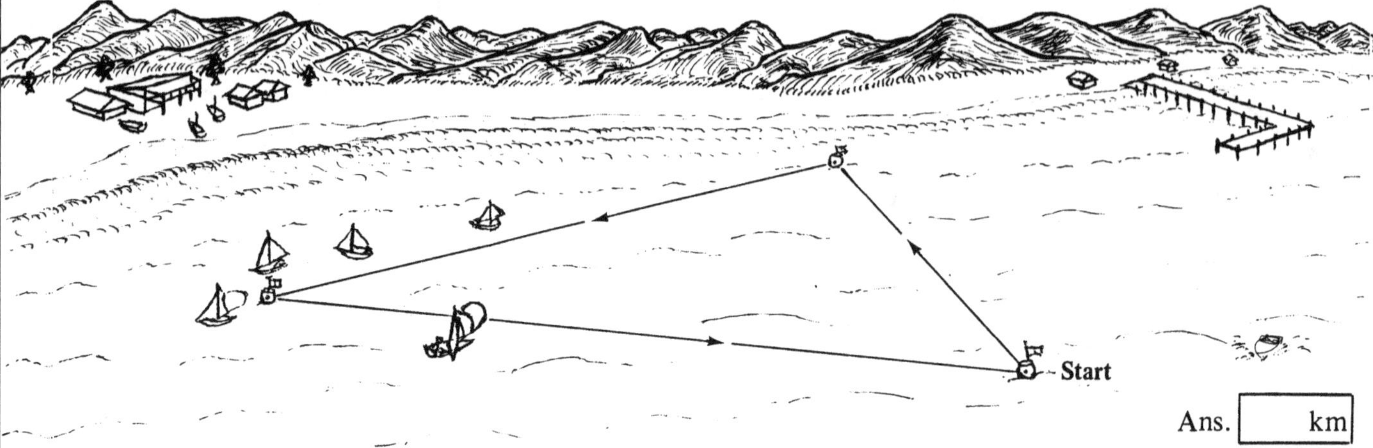

6. If 1 cm is used to represent 500m, find the distance covered in this orienteering course in metres.

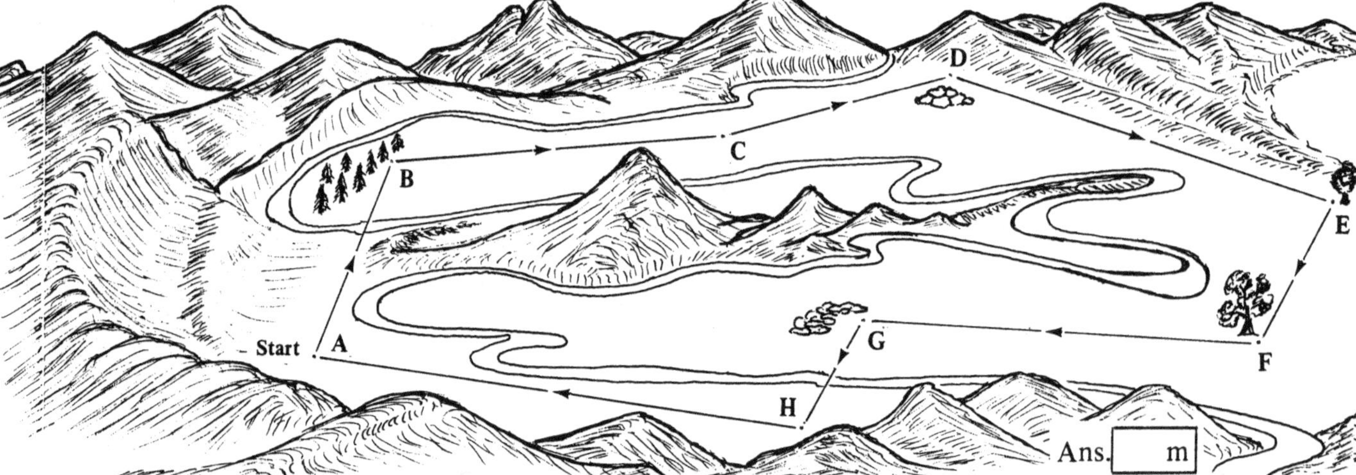

7. If 1 cm is used to represent 100m, find the distance around this obstacle course in metres.

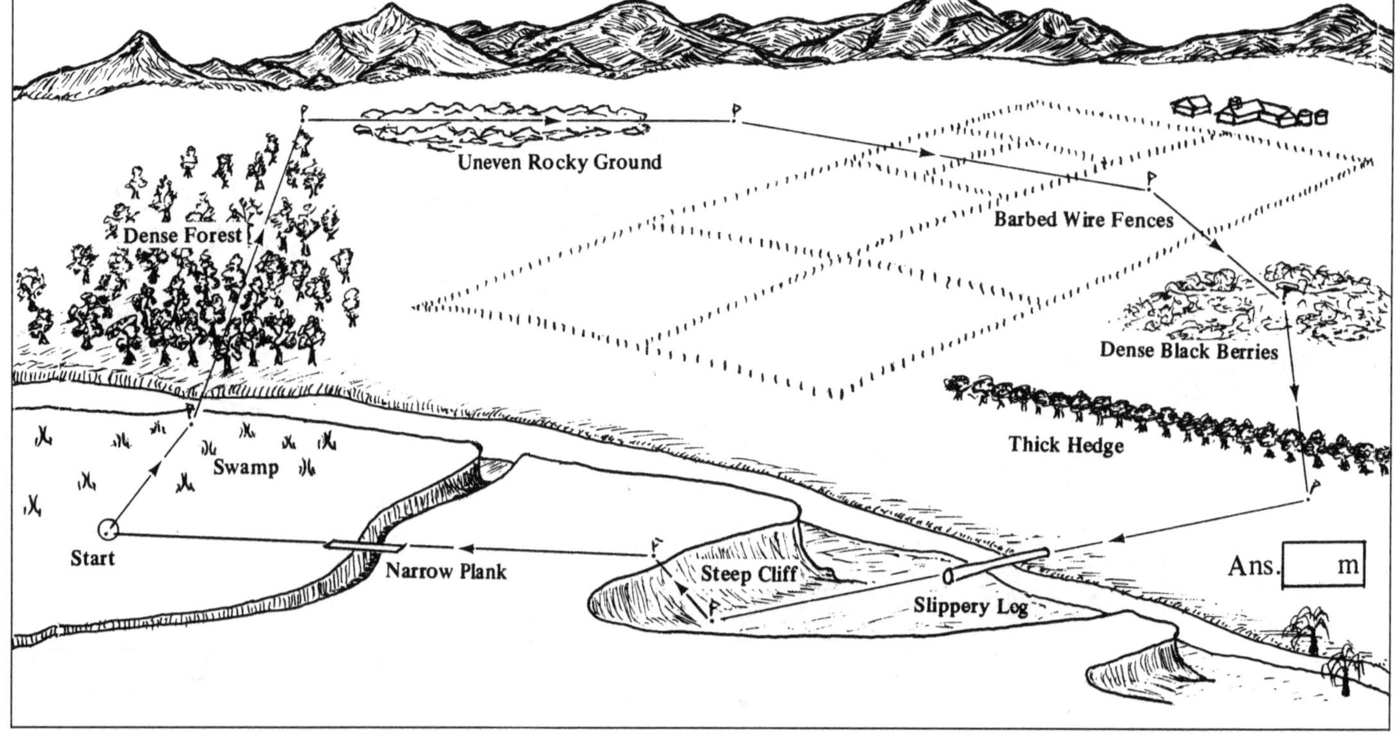

8. If 1 cm is used to represent 80 metres, find the distance around this 9-Hole Golf Course in metres. Do not take into account the distance from each green to the teeing ground

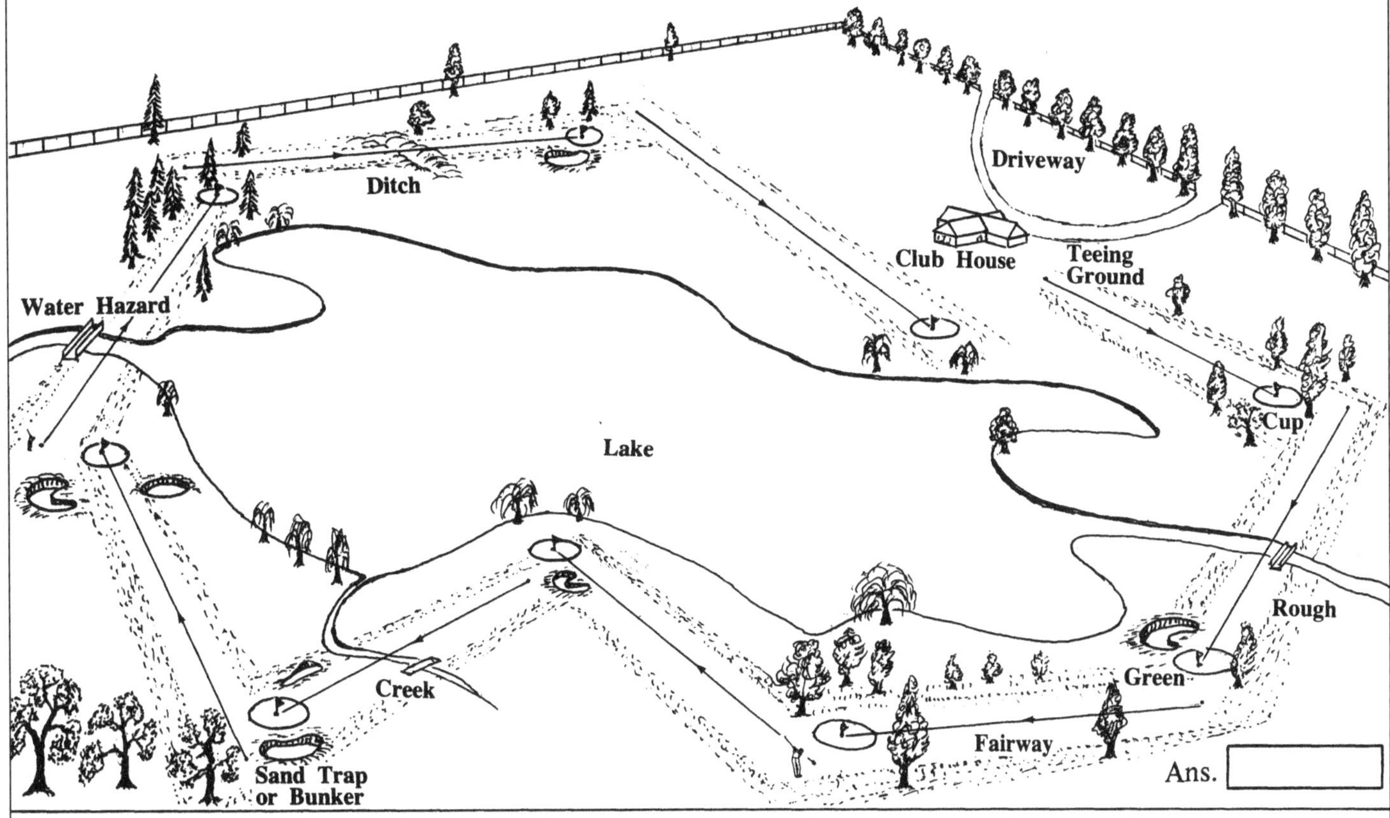

Ans. ____

9. If 1 cm is used to represent 250 metres, find the distance around this bicycle road racing tack in metres

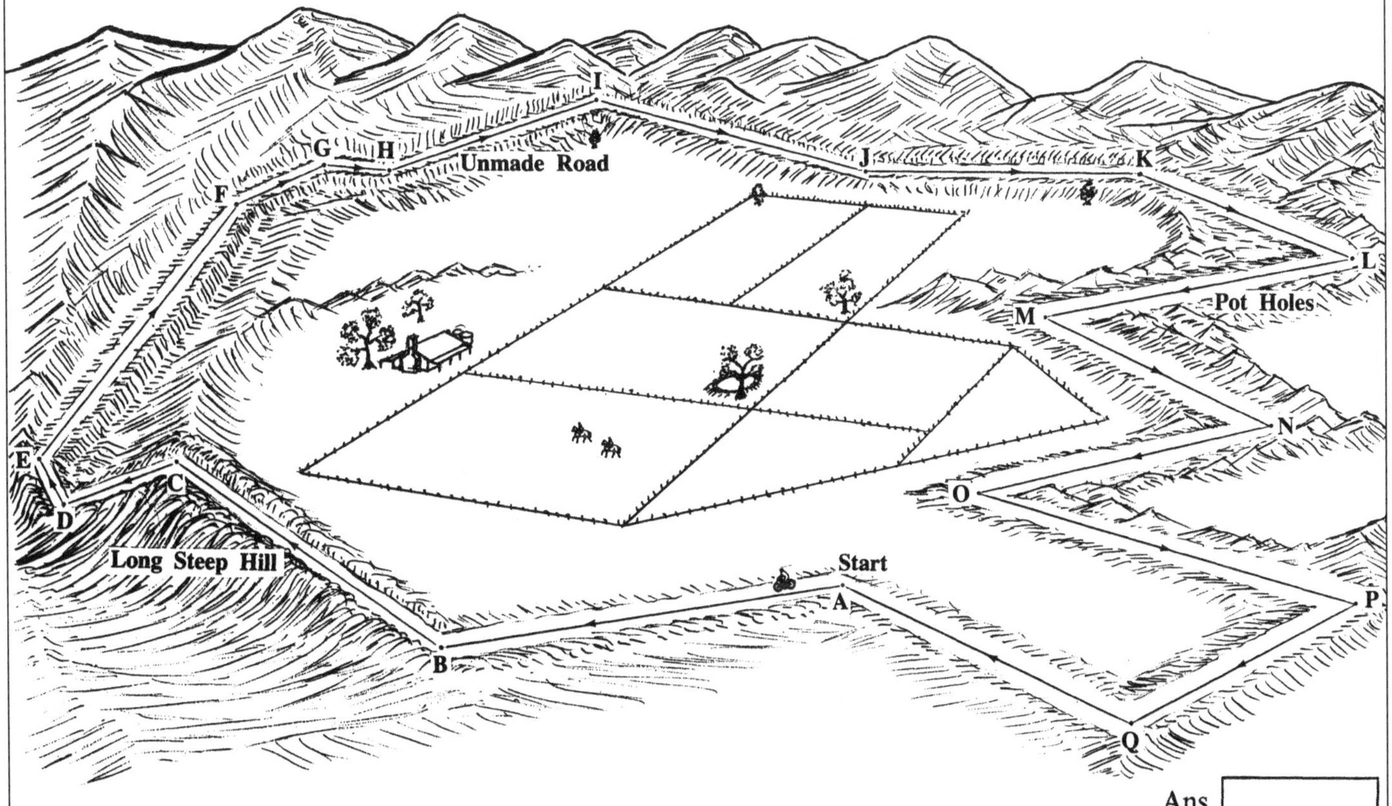

Ans. ____

LINEAR MEASURE

10. If 1 cm is used to represent 2 kilometres, find the distance around this car rally course in kilometres.

Dotted lines represent alternate routes which must not be used. Each car must pass thought the official points marked with capital letters.

Danger
Steep Cliff
Narrow Bridge
Ditch
O
N
Dangerous Bend
P
M
Hair-pin
Bend
Gravel Edges
Rally Point
Start
K
Q
Steep Grade
Narrow Road
L

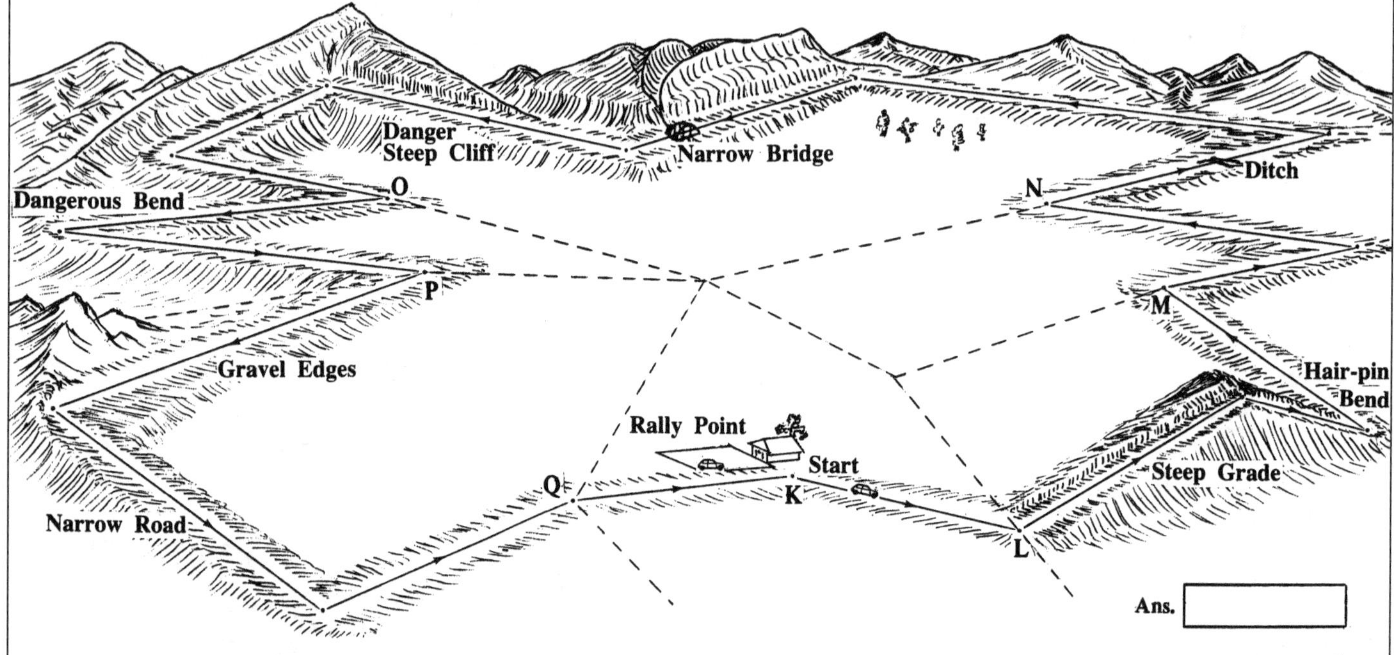

Ans.

11. A Vintage Tiger Moth air race is to be held over the course shown. 1st Prize $30,000. Both the start and finish of the race is to be Moorabbin Airport. If i cm is used to represent 48 kilometres, find the distance around the course in kilometres.

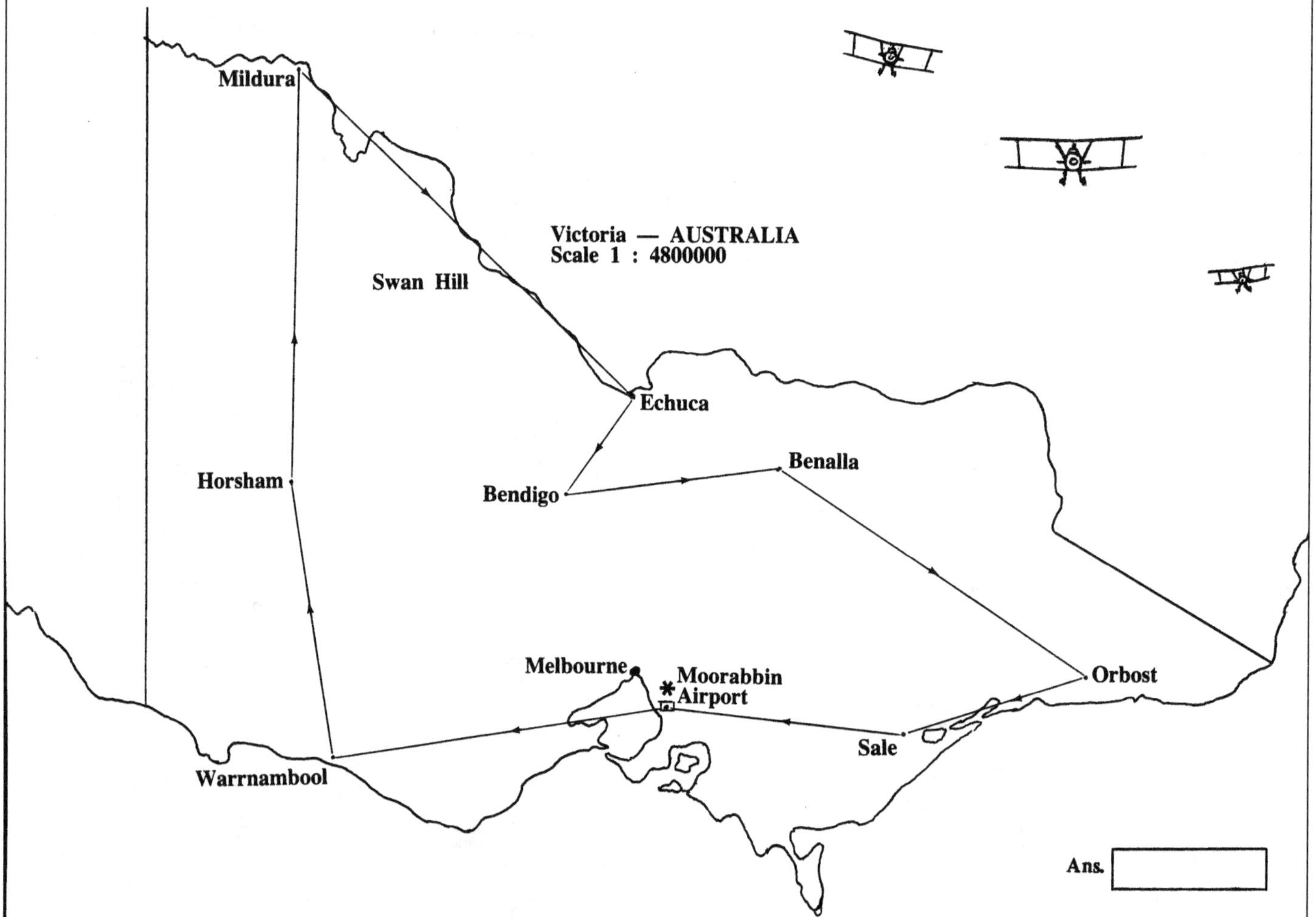

Mildura
Victoria — AUSTRALIA
Scale 1 : 4800000
Swan Hill
Echuca
Benalla
Horsham
Bendigo
Melbourne
Moorabbin
Airport
Orbost
Sale
Warrnambool

Ans.

LINEAR MEASURE

Scale Drawing:

If 1 mm is used in a drawing to represent 1 centimetre, this is referred to as a scale of 1 : 10 (said as 1 to 10)

Example I

State the scale used if 1 mm represents 4 cm.
1 mm represents 40mm
Scale is 1 : 40

Example II

State the scale used if 1 cm represents 5m.
1 cm represents 500 cm
Scale is 1 : 500

In each case state the scale if:

1.

(a)	1mm represents 3cm	Ans.	1 : 30
(b)	1mm represents 7cm	Ans.	
(c)	1mm represents 10cm	Ans.	
(d)	1mm represents 25cm	Ans.	
(e)	1mm represents 78cm	Ans.	

2.

(a)	I cm represents 1m	Ans.	1 : 100
(b)	I cm represents 3m	Ans.	
(c)	I cm represents 8m	Ans.	
(d)	I cm represents 10m	Ans.	
(e)	I cm represents 63m	Ans.	

3.

(a)	1cm represents 1km	Ans.	1 : 100 000
(b)	1cm represents 4km	Ans.	
(c)	1cm represents 10km	Ans.	
(d)	1cm represents 35km	Ans.	
(e)	1cm represents 98km	Ans.	

4.

(a)	I cm represents 15m	Ans.	
(b)	I mm represents 7m	Ans.	
(c)	I cm represents 2km	Ans.	
(d)	I mm represents 13cm	Ans.	
(e)	I cm represents 0.5m	Ans.	

5. Using a scale of 1: 10, what length in mm is represented by:

		Ans.			Ans.
(a)	2cm ?		(f)	2.7cm ?	
(b)	7cm ?		(g)	9mm ?	
(c)	3mm ?		(h)	10cm ?	
(d)	8mm ?		(i)	6.3mm ?	
(e)	1.5cm ?		(j)	13cm ?	

6. Using a scale of 1:100, what length in cm is represented by:

		Ans.			Ans.
(a)	3cm ?		(f)	4.5cm ?	
(b)	8cm ?		(g)	17cm ?	
(c)	4mm ?		(h)	6.4mm ?	
(d)	7mm ?		(i)	8.3cm ?	
(e)	12cm ?		(j)	53mm ?	

7. Using a scale of 1: 1000, what length in metres is represented by:

		Ans.			Ans.
(a)	6cm ?		(f)	56mm ?	
(b)	9mm ?		(g)	12.7cm ?	
(c)	7.5mm ?		(h)	24.5mm ?	
(d)	14cm ?		(i)	32.4cm ?	
(e)	8.3cm ?		(j)	0.85mm ?	

8. Using a scale 1 : 1000 000 what length in km is represented by:

		Ans.			Ans.
(a)	7mm ?		(f)	1m ?	
(b)	3cm ?		(g)	1.2m ?	
(c)	2.6cm ?		(h)	43mm ?	
(d)	8.5mm?		(i)	1.7m ?	
(e)	0.75m ?		(j)	2.69m ?	

Scale Drawing

From the previous calculations it should now be apparent that in scale drawing 1 mm is the basic unit and this is generally used to represent other metric measures. (see table opposite).

Scale	1 mm represents
1 : 10	1 cm
1 : 100	10 cm
1 : 1 000	1 m
1 : 10 000	10 m
1 : 100 000	100 m
1 : 1 000 000	1 km

The following is a map of Australia.
Scale 1 : 20 000 000

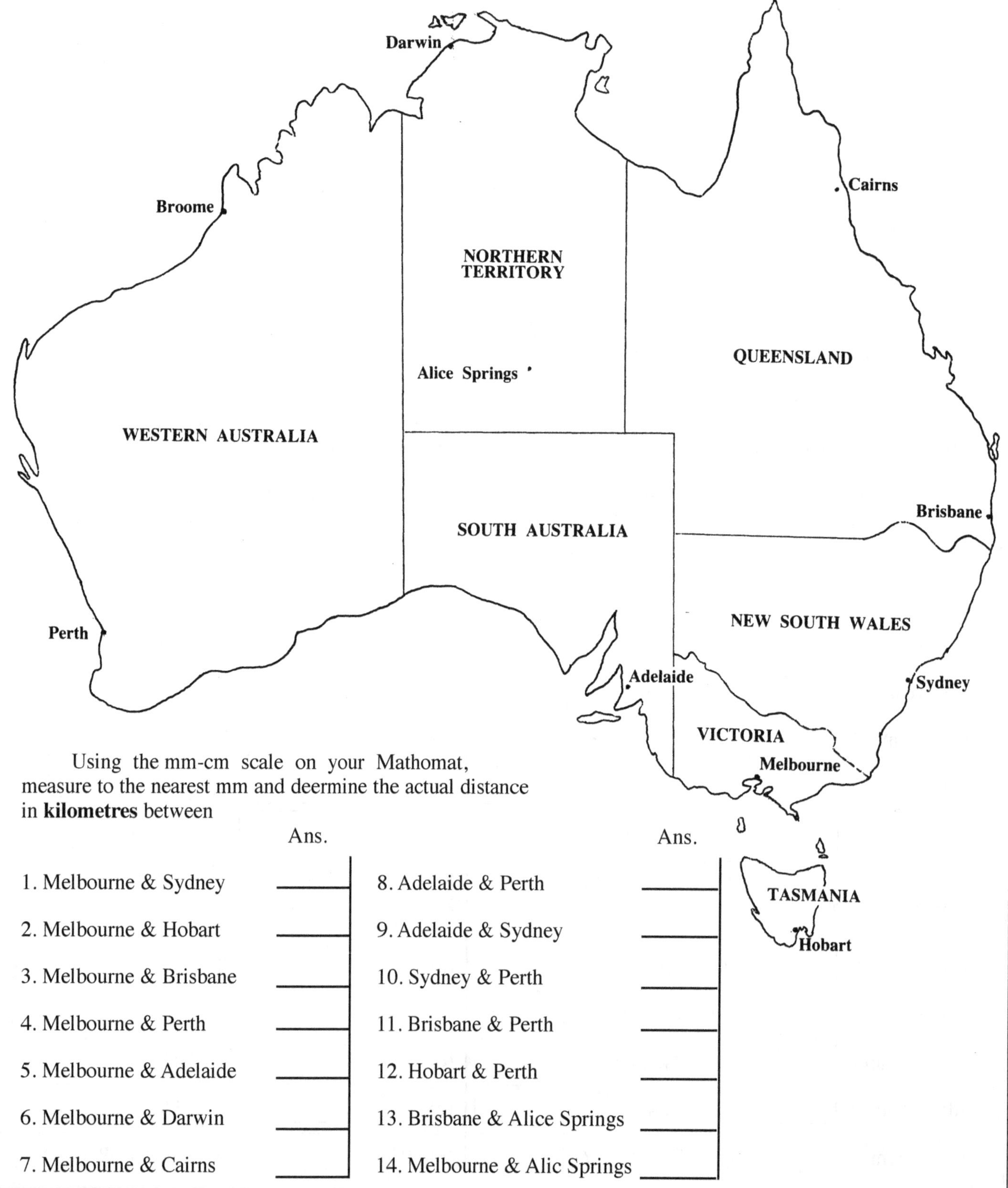

Using the mm-cm scale on your Mathomat, measure to the nearest mm and deermine the actual distance in **kilometres** between

	Ans.		Ans.
1. Melbourne & Sydney	______	8. Adelaide & Perth	______
2. Melbourne & Hobart	______	9. Adelaide & Sydney	______
3. Melbourne & Brisbane	______	10. Sydney & Perth	______
4. Melbourne & Perth	______	11. Brisbane & Perth	______
5. Melbourne & Adelaide	______	12. Hobart & Perth	______
6. Melbourne & Darwin	______	13. Brisbane & Alice Springs	______
7. Melbourne & Cairns	______	14. Melbourne & Alic Springs	______

LINEAR MEASURE

Scale Drawing:

The following represents scale dawing. In each case using the mm-cm scale on your Mathomat and the scale shown, measure to th nearest mm the dimensions marked with capital letters, then calculate then full size dimensions of the object shown. Clerly write your answers in th space provided.

1. Scale 1 : 100

S = m

H = m

T = m

2. Scale 1 : 800

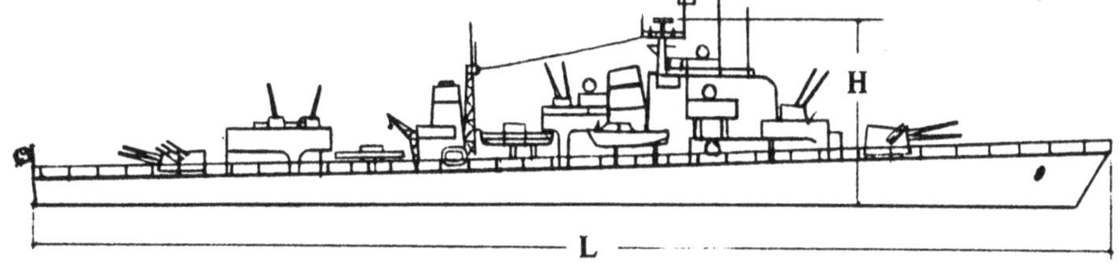

L = m

H = m

3. Scale 1 : 3000

L = m

D = m

4. Scale 1 : 125

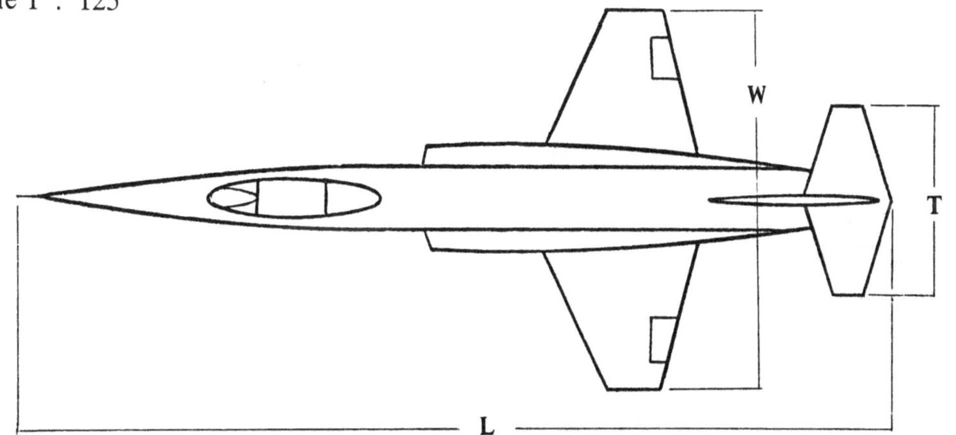

L = m

W = m

T = m

LINEAR MEASURE

Scale Drawing:

5. Scale 1 : 100

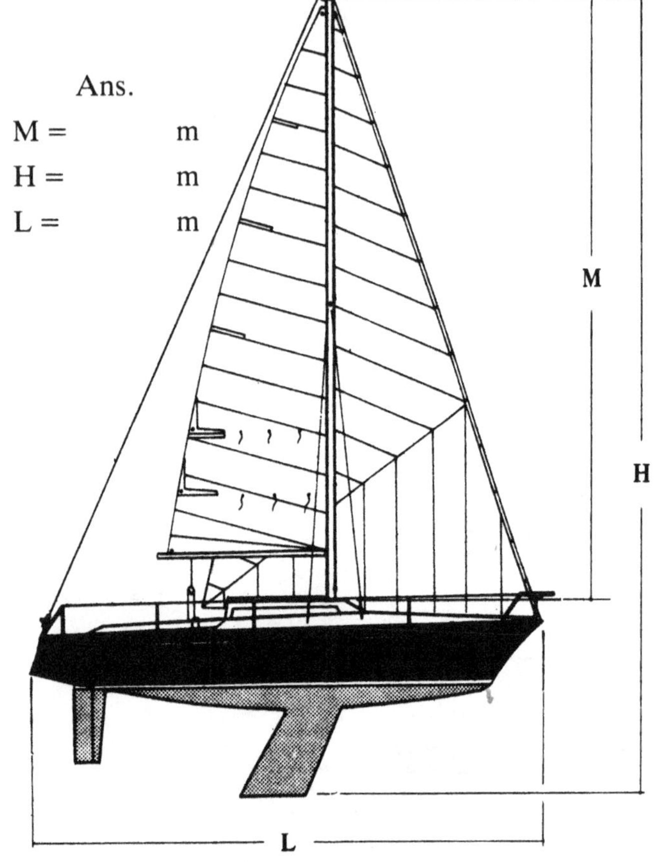

Ans.
M = m
H = m
L = m

6. Scale 1 : 600

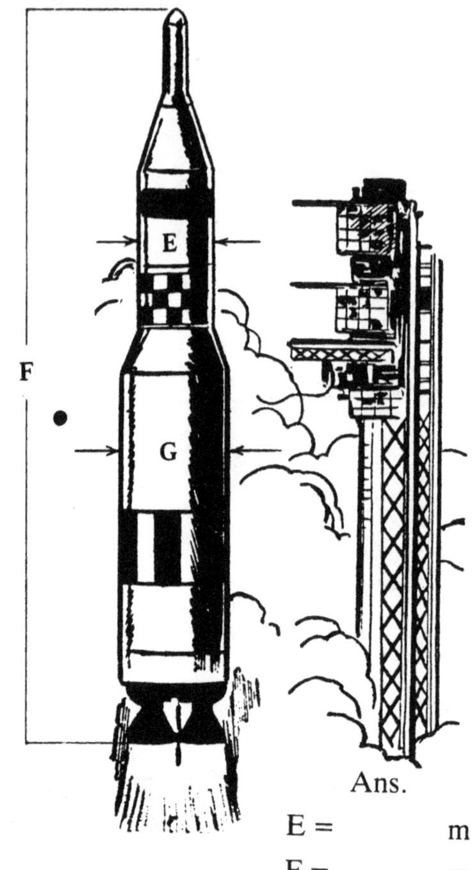

Ans.
E = m
F = m
G = m

7. Scale 1 : 30

Answers: A = m, B = m, C = m, D = m, E = m.

Scale Drawing:

Class or Home Assignment

The following is the plan of a neat, cost efficient 3 bedroom home.

1. Using the scale shown calculate in both mm and m the full size
   - (i) external dimensions (capital letters)
   - (ii) internal dimensions of each room
2. From the external dimensions calculate the **area** of this house in square metres (m^2)

PLAN Scale **1 : 120**

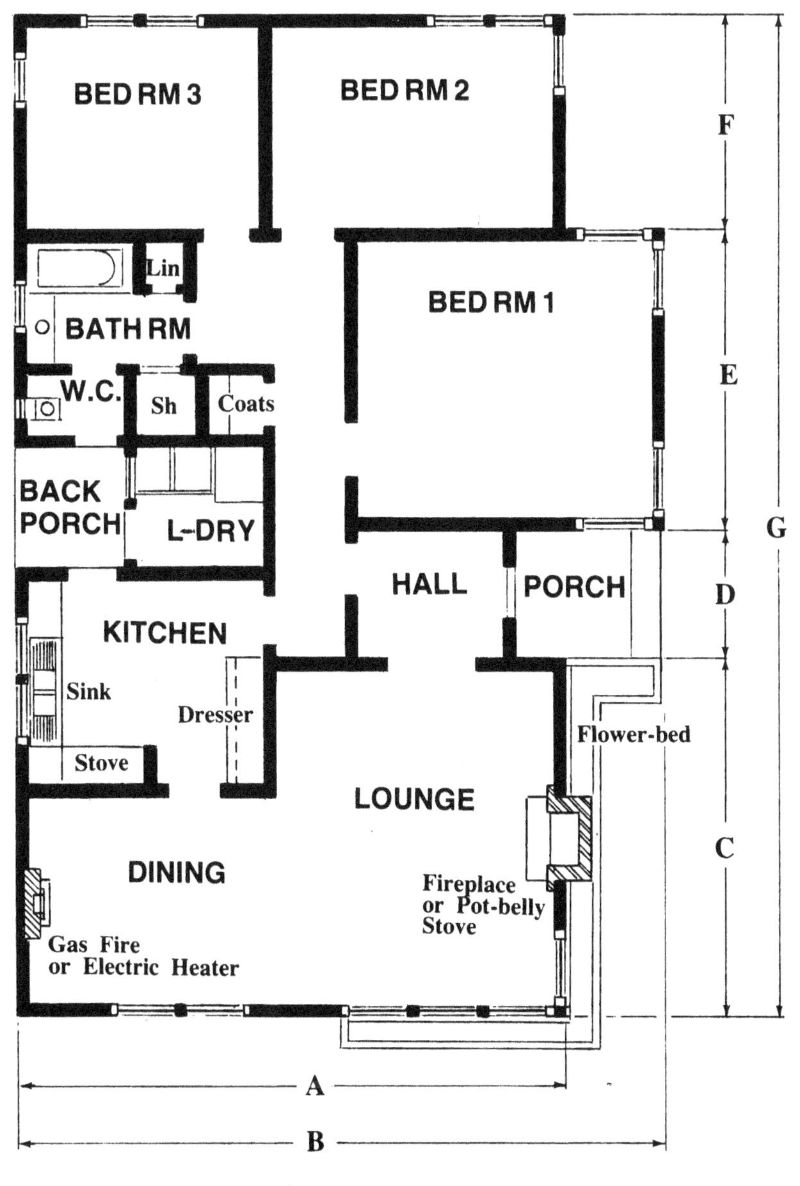

Answers

1. (i)

A =	mm =	m
B =	mm =	m
C =	mm =	m
D =	mm =	m
E =	mm =	m
F =	mm =	m
G =	mm =	m

(ii)

Porch	mm x	mm
	m x	m
Hall	mm x	mm
	m x	m
Lounge	mm x	mm
	m x	m
Dining	mm x	mm
	m x	m
Kitchen	mm x	mm
	m x	m
Bd Rm. 1	mm x	mm
	m x	m
Bd Rm. 2	mm x	mm
	m x	m
Bd Rm. 3	mm x	mm
	m x	m
Bath Rm.	mm x	mm
	m x	m
Laundry	mm x	mm
	m x	m
Back Porch	mm x	mm
	m x	m

2. **Area** of house

= ________ m^2

FLOW CHARTS 1

Before an problem is solved or detailed instructions are given, a certain amount of planning should be done. This generally involves listing all possible steps and then carefully rearranging them in a logical order.

This list of steps to be performed in sequence is called a Logic Diagram or Flow Chart.

Computer programmers and time and motion study experts are continually using these charts as they give a clear and vusual interpretation of the progrm. However, each and everyone of us, at one time or another uses a sequence of steps to perform a particular operation eg: baking a cake, changing a car wheel, eating and drinking, riding a bike, using the telephone, cleaning shoes, calculating income tax, making some toast, driving a car, mowing the lawn.

Much of the planning required is often done unconsciously and without much thought, such that some of the operations we are involved in may be time wasting and inefficient. For this reason and for the fact that it is good training in clear thinking, some flow charts of everyday actions will be included. For clarity, the steps in a flow chart may listed within squares, circles or other geometric figures connected by arrows designating the flow or direction of the process.

Example: Here is a simple flow chart for getting up in the morning.

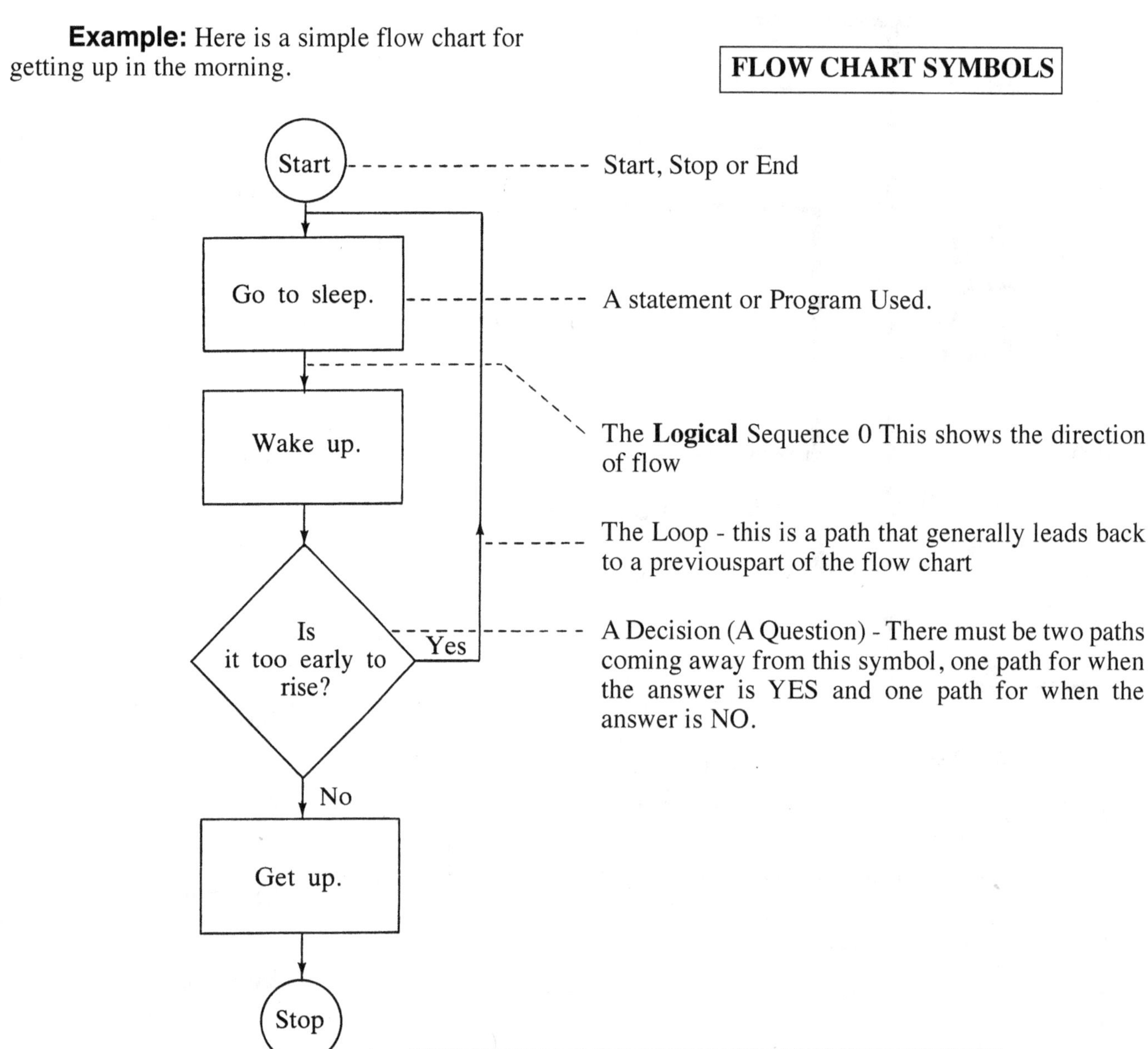

Mathomat is particularly useful here for quickly drawing the geometrical shapes used.

Note:

The following flow charts
(a) do not necessarily show all the steps that may be involved in a particular action.
and (b) may be constructed differently to that shown.

However, use the steps shown to construct and number the resulting flow charts.

Example: Waiting at A Railway Station to Board A Train

Program

1. Start
2. Is a traion coming? (Yes) — (No) 3. Look for next train
4. Note destination on front of train
5. Is it the train I want? (Yes) — (No) 6. Look for next train.
7. Has the train stopped? (Yes) — (No) 8. Wait until train stops
9. Move to carriage opposite a door.
10. Is the door open? (Yes) — (No) 11. Open door
12. Step aboard.
13. Close door.
14 Stop

Flow Chart

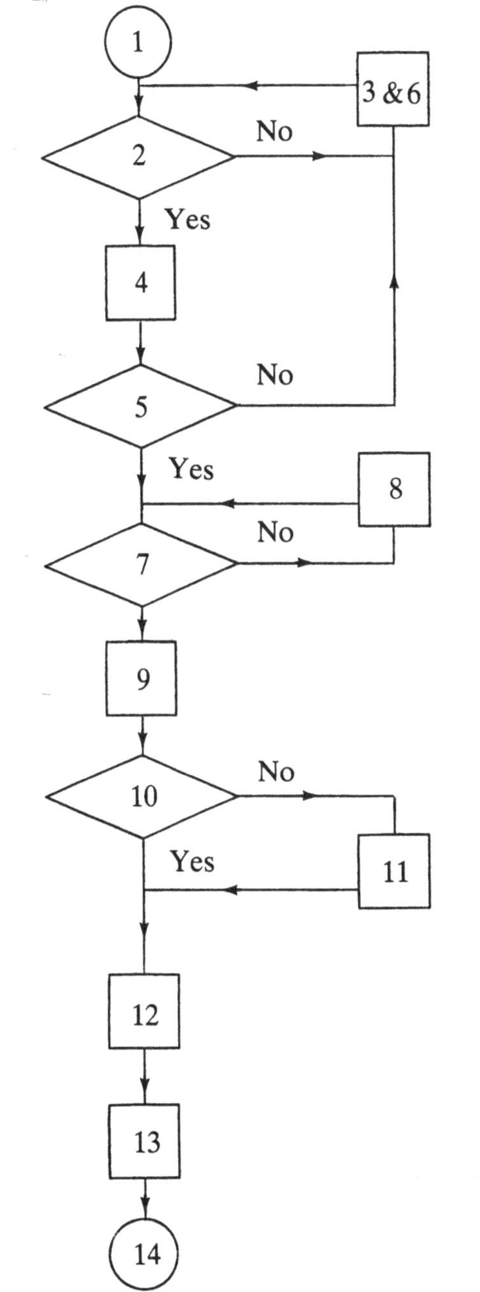

Questions on the above flow chart
(a) How many possible loops involved?
(b) How many decisions?
(c) How many possible statements?

Answers

FLOW CHARTS

In the following examples the program steps have been written in a logical sequence. Using Figs. 14, 5 and 15 of your Mathomat draw thee flow charts involved in a similar manner to the prvios example. Remeber to number each part of the flow chart and o clearly show the direction of flow.

A. Lighting A Bunsen Burner.

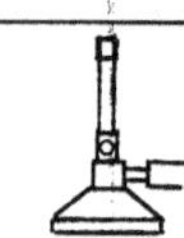

Program

1. Start

2. Connect burner to gas tap.

(No)
4. Close air regulating hole

3. Is air regulating hole closed?
(Yes)

5. Light a match

6. Turn on gas.

7. Light gas.

8. Is a hot flame required?
(No)

(Yes)
10. Open air regulating hole

9. Adjust height of flame.

11. Stop.

Flow Chart

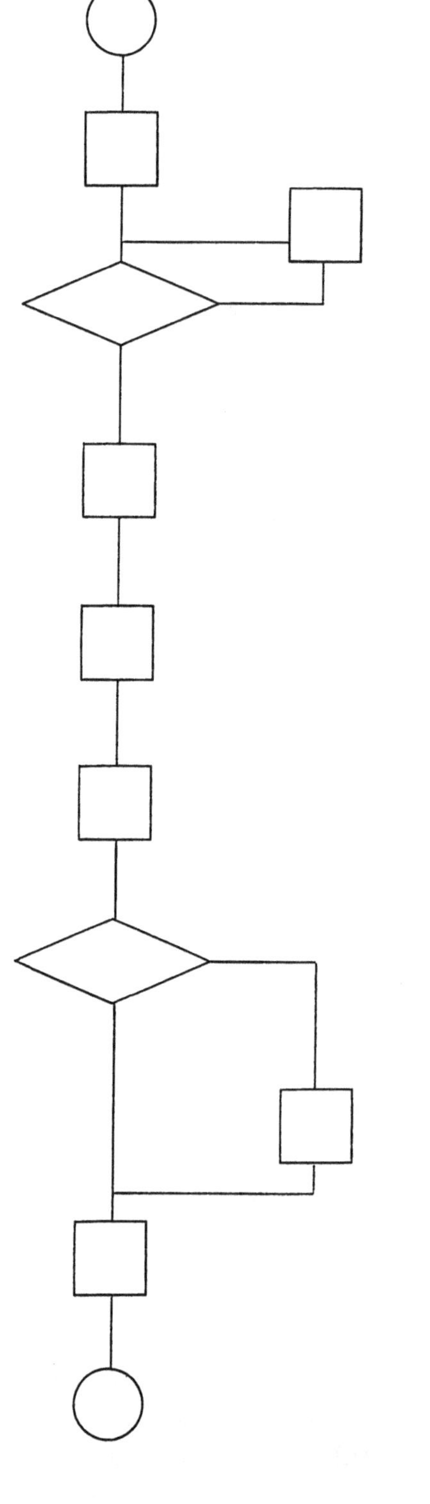

B. Deciding Whether A Number Is Division By 21

Program

Flow Chart

1. Start

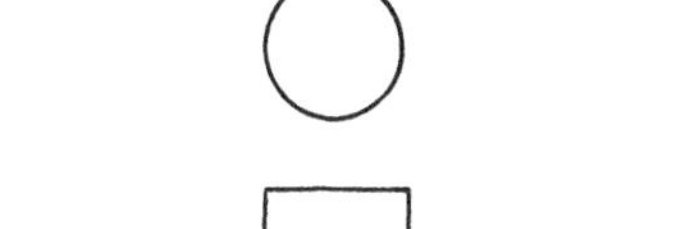

2. Let "N" be the number.

3. Is "N" divisible by 3 and by 7?
   (Yes)

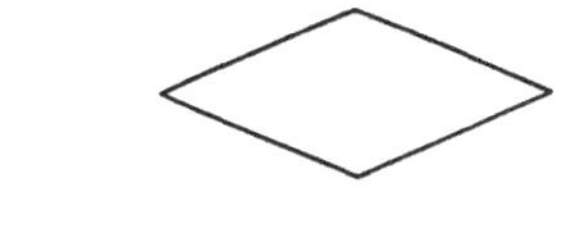

4. "N" is divisible by 21.

(No)
5. "N" is not divisible by 21.

6. Stop

C. Deciding Whether A Scalene Triangle Is Right Angled, Acute Angled Or Obtuse Angled. (Given only the lengths of the 3 sides)

Program

Flow Chart

1. Start

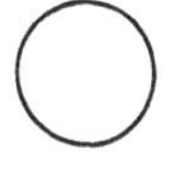

2. Let "C" be the measure of the longest side, and "a" and "b" the measures of the other two sides.

3. Work out the respective values of C2, a2 and b2.

4. Calculate a2 + b2

5. In a2 + b2 = C2? (No)

7. Is a2 + b2 > C2? (No)

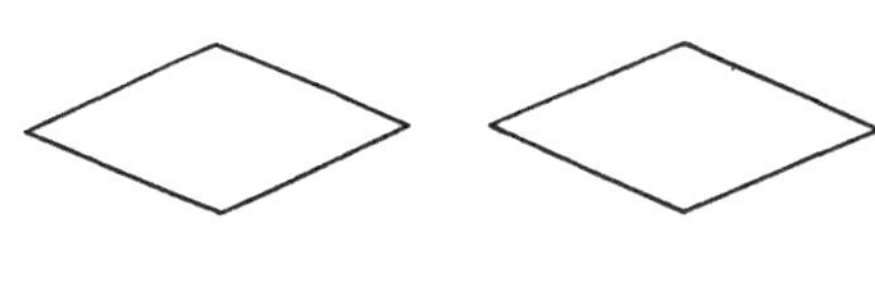

(Yes)
6. Triangle is Right Angled.

(Yes)
8. Triangle is Acute Angled.

9. triangle is Obtuse Angled.

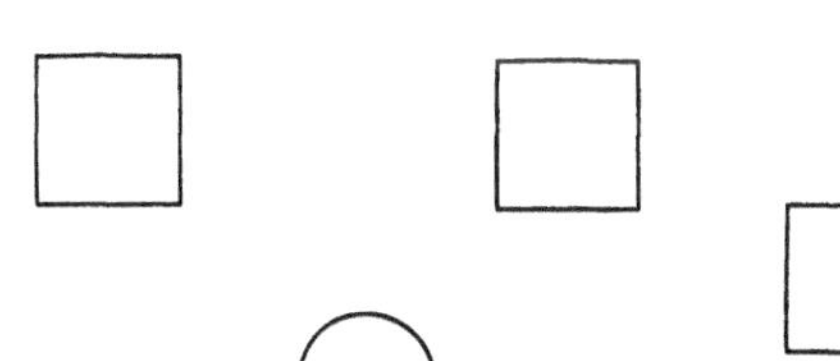

10. Stop

D. Preparation Of A "Medium" Boiled Egg.

Program		Flow Chart
1. Start		
2. Place enough water in a saucepan to cover egg		
3. Switch on stove element		
4. Place saucepan on stove element.		
5. Pick up egg.		
6. Pierce the shell of the "blunt" end of the egg with a small hole.	(No) 8. Wait until water boils.	
7. Is water in saucepan boiling (Yes)		
9. Pick up a spoon large enough to hold egg.		
10. Place egg on spoon.		
11. Lower egg slowly into boiling water		
12. Put dwon sppon.		
13. Is egg large? (Yes)	(No) 14. Leave in boiling water for approx. 31/2 to 5 minutes.	
15. Leave in boiling water for approx 5 to 7 minutes.		
16. Pick up spoon.		
17. Life egg out.		
18. Place in egg cup.		
19. Put down spoon.		
20. Srtop.		

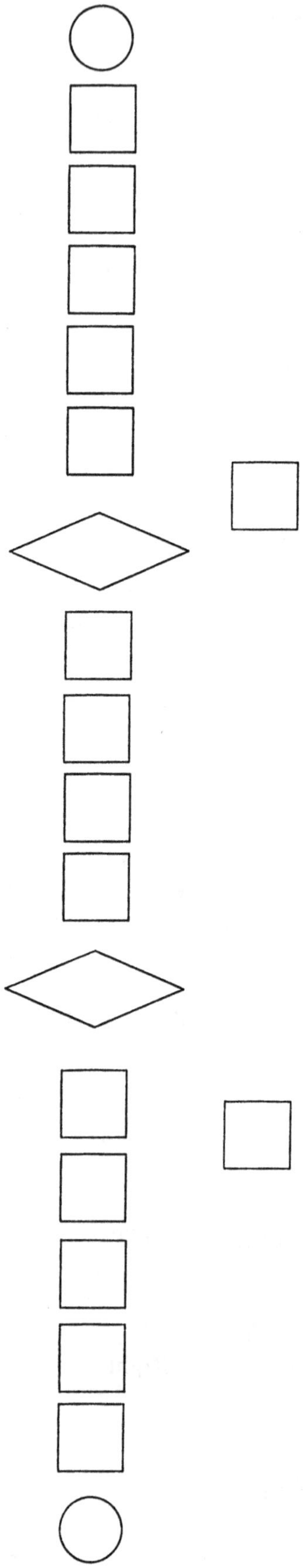

E. Starting A Car In Winter

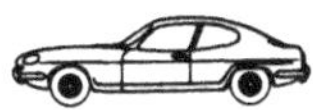

If your car has to be left ouvernight in wet or misty wearther you may find difficulty starting in the morning. This is due to damp, so dont go on grinding away hopefully with the starter, simply complete the steps in the program below.

Program	Flow Chart
1. Start	
2. Lift up bonnet.	
3. Undo distributor cap securing straps.	
4. Lift off distributor cap.	
5. Use a clean dry cloth to wipe dry both the inside and the outside of the cap.	
(No) 7. Obtain a spray can of water repellent light oil.	
6. Have you a spray can of water repellent light oil? (Yes)	
8. Spray inside the cap and 'points' with water repellent oil.	
9. Replace cap on distributor.	
10. Secure cap with straps	
11. Closebonnet.	
12. Start Car.	
13. Stop.	

F. To Switch On And View A Television Show.

Program	Flow Chart
1. Start.	
(No) 3. Plug set in.	
2. Is set plugged in? (Yes)	
4. Switch set on.	
(No) 6. Select correct channel.	
5. Is set on correct channel? (Yes)	
7. Adjust sound and colour?	
8. Sit down.	
(No) 10. Wait for 'show' to commence	
9. Is 'show' commencing? (Yes)	
11. View 'show'.	
12. Get up when 'show' is finished.	
13. Switch off set.	
14. Stop.	

G. To Remove A Motor Registration Certificate Transfer From A Car Window.

(The following program is very effective as usually the transfer can be removed easily without being damaged).

Program		Flow Chart
1. Start		
	(No) 3. Obtain the new transfer	
2. have you the new transfer? (Yes)		
4. Use a sponge dipped in water to thoroughly wet the back of the old transfer.		
5. have you a piece of clear plastic wrap sufficient to cover the transfer? (Yes)	(No) 6. Obtain a piece of clear plastic food wrap sufficient to cover the transfer	
7. Use the sponge to wet one side of this plastic.		
8. Place the wet side of the plastic over the back of the old transfer.		
9. Leave for 5 minutes		
10. Have you a thin razor blade or knife (Yes)	(No) 11. Obtain a thin razor blade or knife	
12. Remove plastic food wrap.		
13. Use razor blade to peel off the old transfer.		
14. Stop.		

H. To Give Eye Drops To The Right Eye Of A Patient.

(A particularly good metod for young children.)

Program		Flow Chart
1. Start.		
2. Sit patient down in a chair, relaxed.		
	(No) 4. Obtain correctt eye drops.	
3. Have you obtained the correct eye drops? (Yes)		
	(No) 6. Carefully read the directions on the label.	
5. Have you carefully read the directions on the label? (Yes)		
7. Move behind the patient.		
8. Ask the patient to tilt their head back and look at the ceiling directly above them.		
9. Ask the ptient to close their eyes and to tilt their head slightly to the right.		
10. Place the required number of drops in the inner corner of the right eye.		
11. Ask the patient to open their eyes and to blink a number of times.		
12, Stop.		

FLOW CHARTS

10

I. Removing A Car Wheel.

Program		Flow Chart
1. Start.		
	(No) 3. Plac car in gear or park with hand brake on.	
2. Is car in gear or park with hand brake on? (Yes)		
4. Obtain a Jack, Wheel Chocks and Wheel Brace.		
5. Place wheel chocks at front and rear wheels diagonally opposite wheel to be removed.		
6. Remove wheel cap.		
7. Use wheel brace to loosen wheel nuts one turn.		
	(No) 9. Place 'jack' in correct postion	
8. Has the 'jack' been placed in the correct postion? (Yes)		
10. Jack up the car so that wheel is off the ground.		
11. Remove wheel nuts.		
12. Place in hub cap.		
13. Lift off wheel.		
14. Stop.		

J. To Make A Batch Of Golden Brown Plain Scones.

(Using a 375g NET Packet off Scone Mix)

Program		Flow Chart
1. Start.		
	(No) 3. Heat oven to hot.	
2. Has oven been preheated to hot? - Gas 230•C, Electric 260•C. (Yes)		
4. Empty 1 packet of Scone Mix into a bowl.		
5. Add 1 cup of milk.		
6. Mix to a moist light dough		
	(No) 8. Prepare a floured board.	
7. Has a floured board been prepared? (Yes)		
9. Place dough on floured board		
10 Knead dough lightly for a few seconds.		
11. Roll out to approx. 15 to 20mm thickness		
12. Cut dough pats to size required	(No) 14. Prepare greased tray.	
13. Has a lightly greased tray been prepared? (Yes)		
15. Place pats on greased tray.		
16. Place greased tray and pats in hot oven.		
17. Bake for approx. 13 to 16 m inutes.		
18. Stop.		

K. To Stop A Sudden Nosebleed (Epistaxis)

Program		Flow Chart
1. Start.		
2. Sit upright relaxed and leaning forward.		
3. Firmly pinch-off the soft front part of the nose between thumb and forefinger.		
4. Breathe through the mouth don't blow your nose.		
5. Maintain sufficient pressure to stop the bleeding.	(No) 7. Maintain pressure.	
6. Has 10 minutes elapsed? (Yes)		
8. Release pressure slowly	(No) Move back to 2.	
9. Has bleeding stopped? (Yes)	(No) 10. Have previous steps been repeated twice? (Yes)	
	11. Reapply pressure on nose.	
	12. Immediately go to a hospital or see a doctor.	
13. Stop.		

FLOW CHARTS

The steps in the following programs have not been given in a logical sequence.

In each case number the steps in a logical sequence and raw the resulting flow chart in the space provided. Use Figs. 14, 5 and 15 of your Mathomat.

A. Making A Cup of Hot Soup
(Using a Cup-Of-Soup Paclet)

Program	Flow Chart
Boil somer water.	
Stop.	
Fill cup with boiling water.	
Stir for 30 to 40 seconds	
Start.	
Empty contents of packet into a suitable cup.	

B. Cleaning and Polishing A Pair Of Leather Shoes.

Program	Flow Chart
Finish polishing with soft cloth.	
Stop.	
Polish with a soft brush.	
Wash hands.	
Strt.	
Brush or scrape off mud.	
Apply polish with a brush.	

C. Striking A Match.

Program	Flow Chart
Strike Match.	
Pick up matchbox.	
Start.	
Close matchbox.	
Does match light?	
Stop.	
Remove a live match.	
Open matchbox.	

D. Making A Pot Of Hot Tea. (Using Tea Pot Bags)

Program	Flow Chart
Allow to "draw" for a few minutes.	
Stop.	
Boil some water.	
Place required number of tea bags in tea pot.	
Start.	
Is teapot empty.	
Pour sufficient boiling water into teapot.	
Empty teapot.	

E. To Pepair A Burnt-out Fuse Wire In A House Fuse Board.

Program	Flow Chart
Remove remaining pieces of fuse wire.	
Stop.	
Turn on main light and power switch.	
Has fuse board been located?	
Replace fuse holder in fuse board.	
Start.	
Re0wire fuse holder.	
Have you a piece of fuse wire of the correct amperage?	
Locate fuse board.	
Has burnt-out fuse holder been located?	
Turn off main light and power switch.	
Obtain a piece of fuse wire of the correct amperage.	
Locate burnt-out fuse holder.	

F. Deciding Whether A Number Is Divisible By 3.

Program	Flow Chart
Let 'N' be the number.	
Stop.	
Is 'S' divisible by 3?	
Start.	
'N' is divisible by 3.	
Let 'S' be the sum of the digits of 'N'.	
'N' is not divisible by 3.	

Now make up logical programs and draw the low charts for the following.

1. Starting an automatic car.
2. Cleaning your teeth.
3. Washing an dpolishing the 'family' car.
4. Paying an electricity account by posting a cheque.
5. Making a local phone call (Metropolitan Area).
6. Mixing black and white paint to obtain the 'right' shade or grey.
7. Replacing a defective light globe in a lamp.
8. Using a pair of 'jump leads' to start a car that has a flat battery.

RECOGNITION OF BASIC SHAPES (Pattern – Symmetry – Colour). Furth Examples 1

The following patterns have been drawn using one or more of the basic shapes in your Mathomat. Redraw these patterns in your excercise book or folio as accurately and as neatly as you can. Use **coloured pencils** to lightly colour-in each pattern.

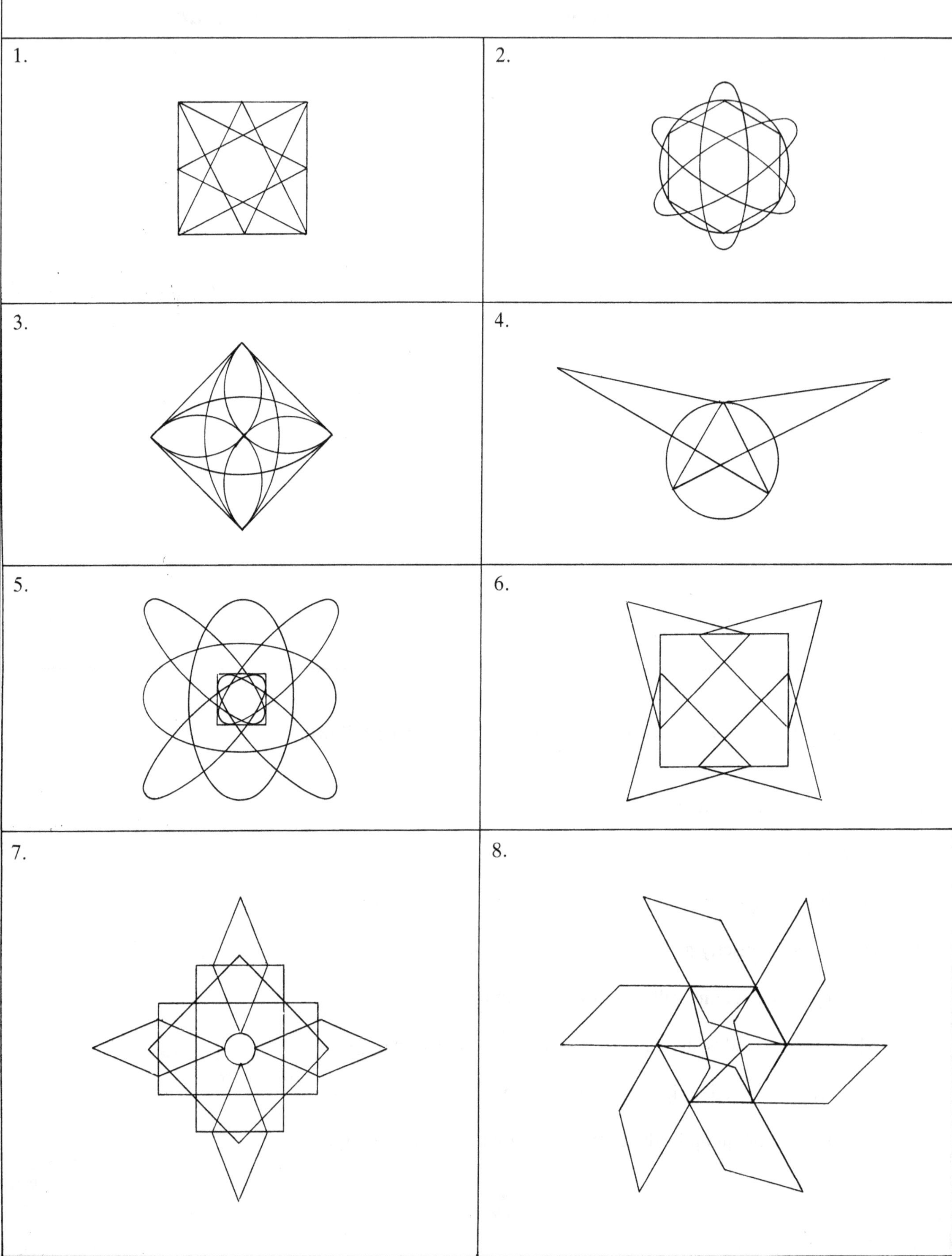

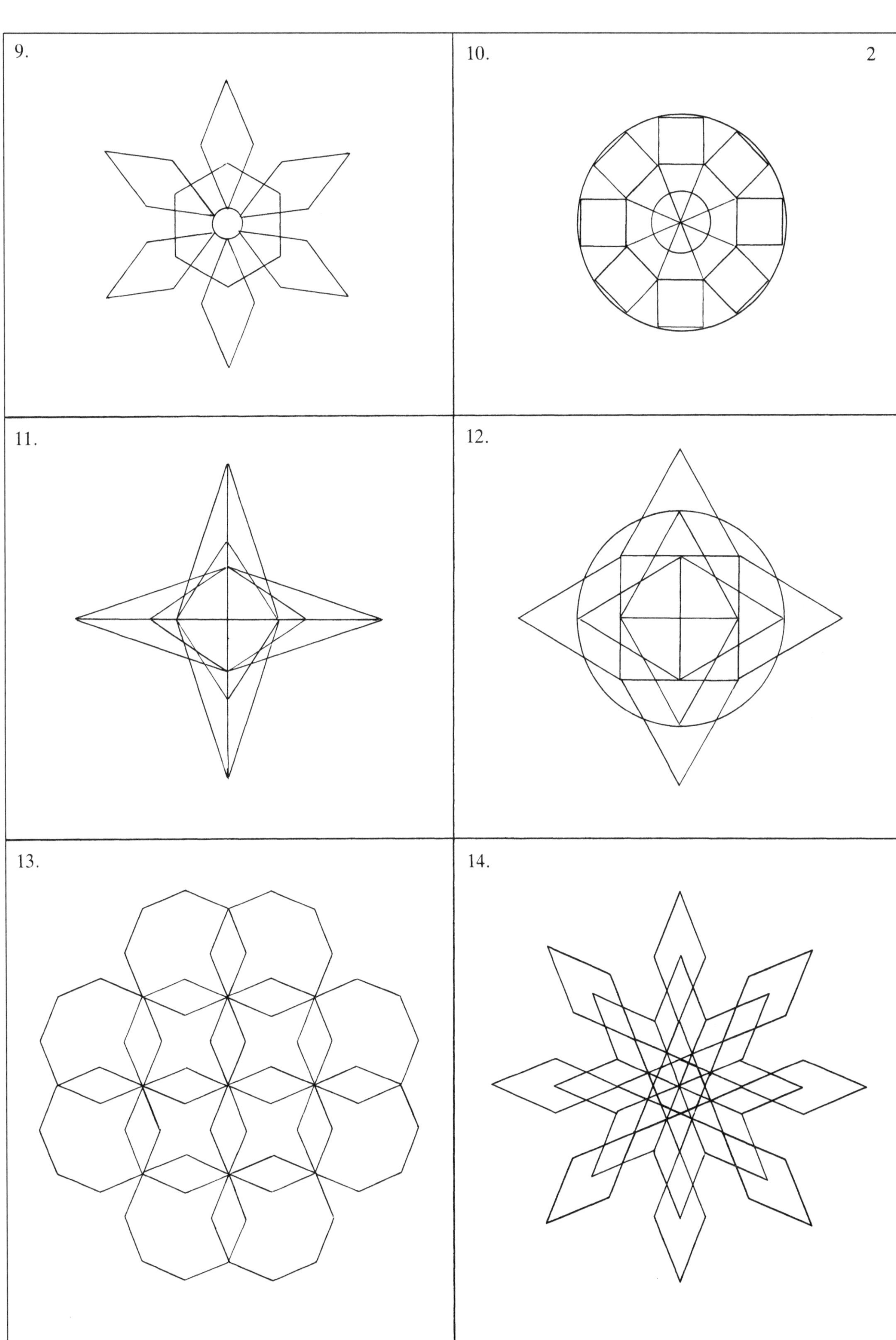
9.
10.
2
11.
12.
13.
14.

ANGLES

1

Why is it Necessary to Study Angles?

Engineers, architects, draftsmen, surveyors, navigators and artists are continually drawing and measuring angles.

The following represents som examples of their work.:

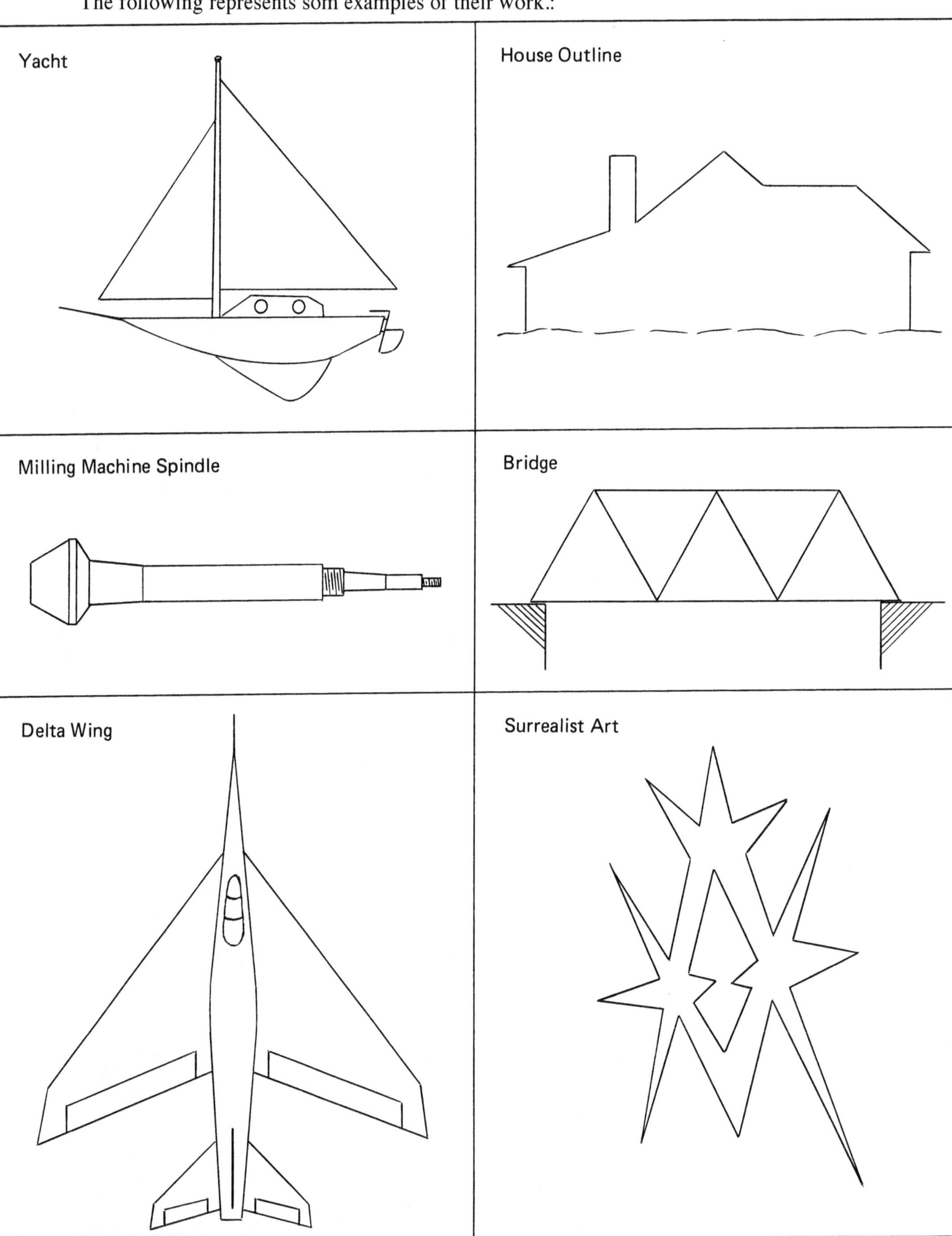

Clearly indicate, by marking in (as shown in the triangle) all corners or angles that are within the above diagrams. Suggestion: Use a red pencil or a red biro.

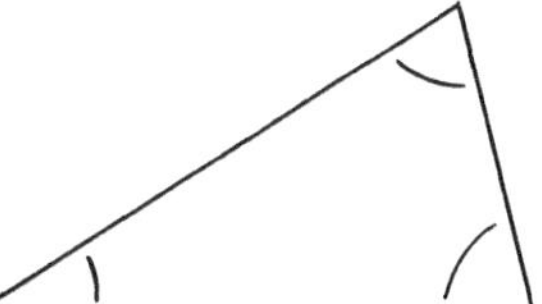

2

As with a line, ray or line segment, an angle is only an idea in our minds,

Consdier now two rays PY and PX, whose end points coincide at P.

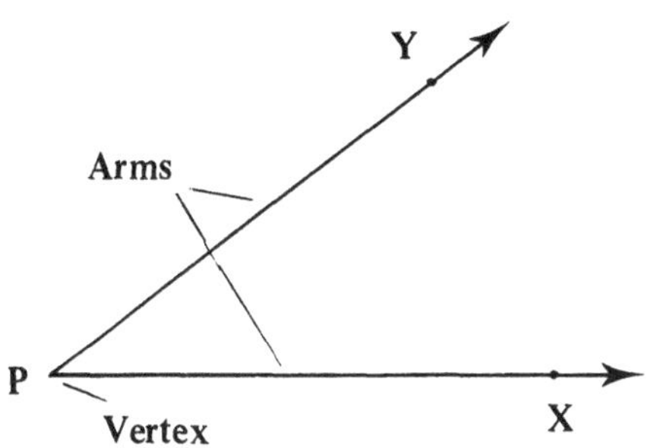

Immediately a geometrical figure has been formed called an **ANGLE.**

The common end point P is called the VERTEX and the two rays the arms of the angle.

In practice the arms of an angle are generally line segments.

Note: In these notes, a line will be considered to be a line segment unless otherwise stated.

An Angle is the union of two lines with a common end point.

Name this angle

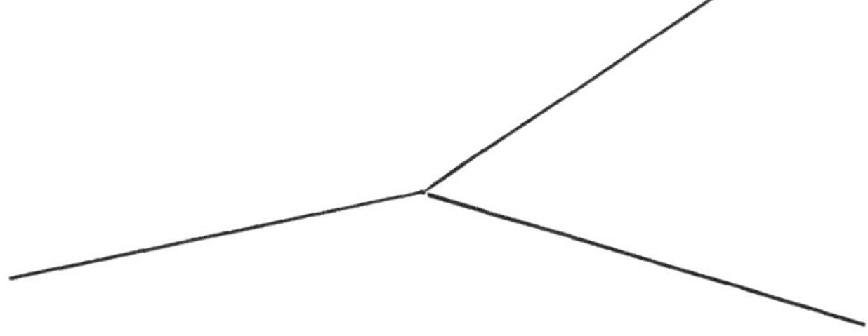

Before an answer can be given to this request, it is obvious that the vertex and the end of the arms must have reference marks. In general, capital letters are used for these points.

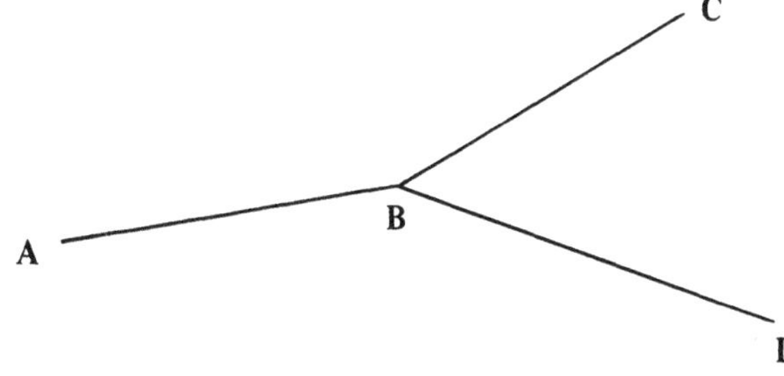

Now when the above request is again asked, to reply can be immediately "which angle?" This eliminates any uncertainty as to the angle under consideration.

The three angles represented here are:

angle ABC, angle CBD and angle ABD

or $\measuredangle$ ABC $\measuredangle$ CBD $\measuredangle$ ABD

1. In a similar manner to the worked example, name all the angles in the following plane figures. **3**

Example: Ans:

N

M P

∡ MNP

∡ NMP

∡ MPN

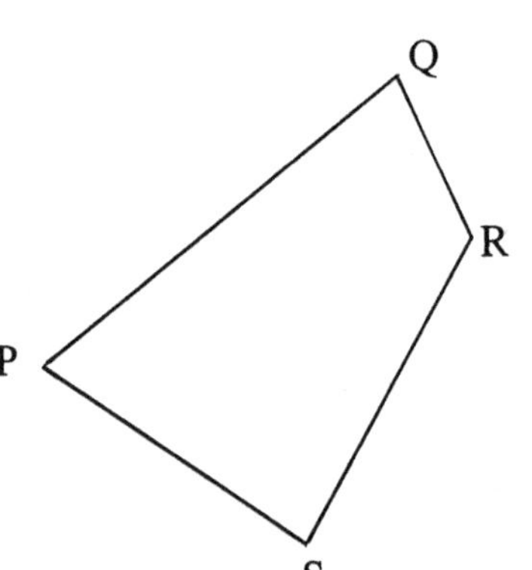

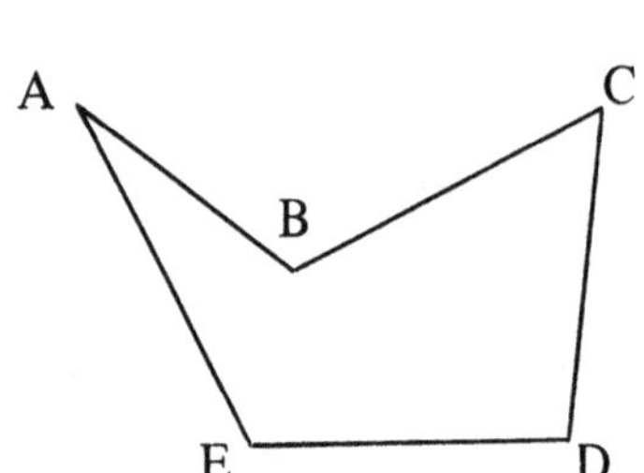

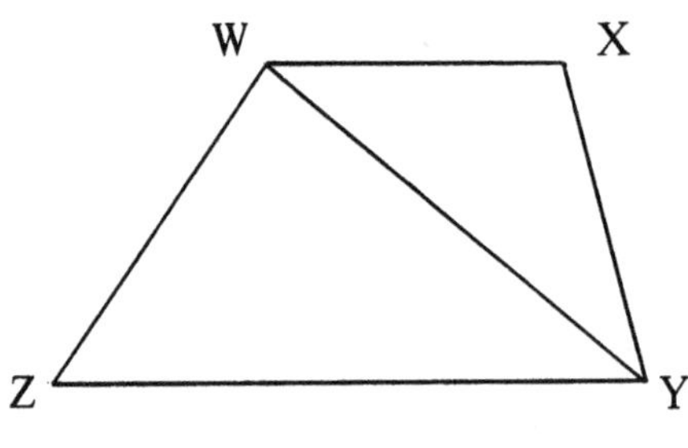

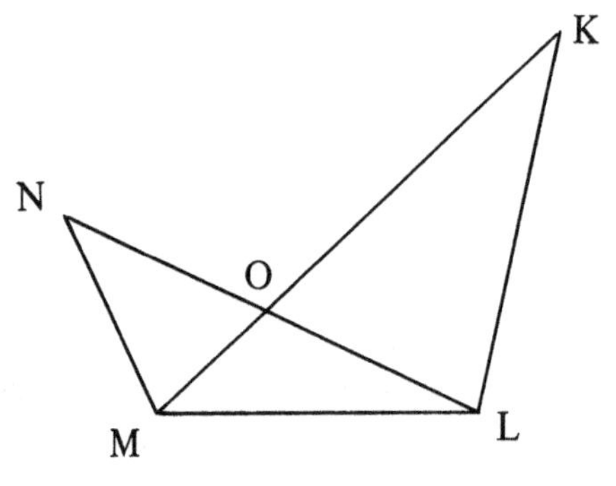

2. Name all the angles in the following 3-dimensional figures.

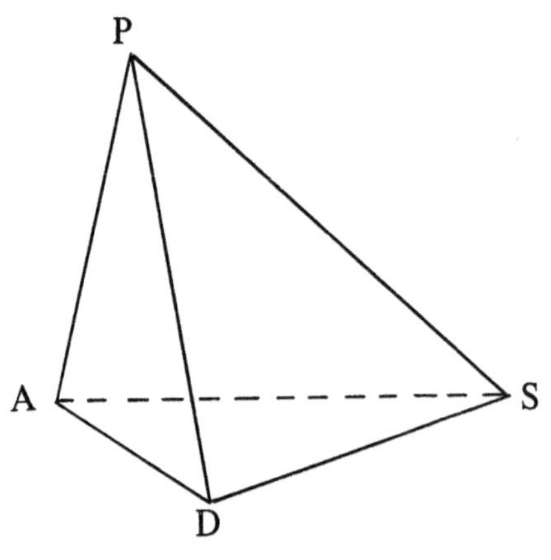

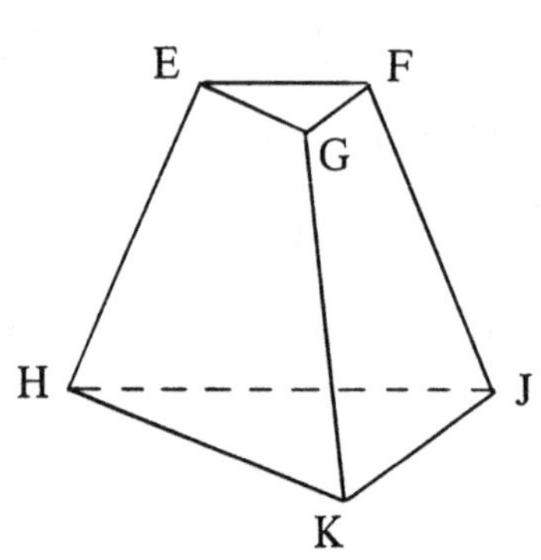

Consider now the following angles. **4**

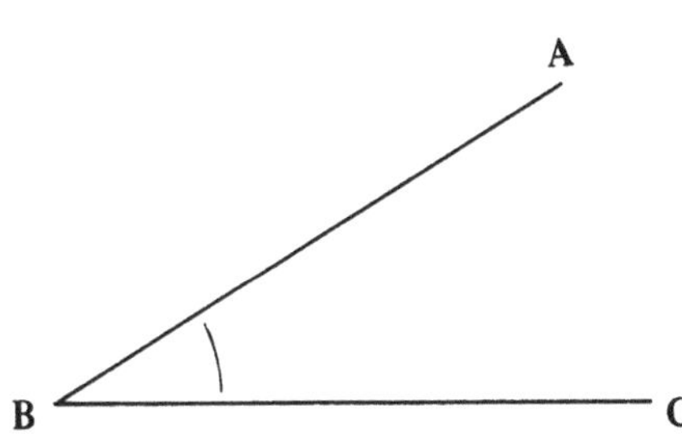

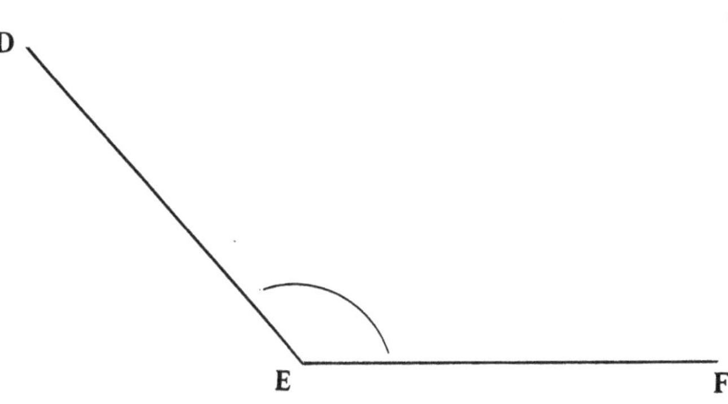

Questions:

1. Do the lengths of the arms of these angles differ? Ans:

2. Does each angle have a vertex? Ans:

3. Are these angles different in some way? Ans:

4. In what way are they different?

   Ans:

Deonstrate to yourself with two rulers or pencils held together at one end, or a blackboard compass, that angles differ from one another by the amount of opening (or turning) between the arms.

The magnitude of an angle is the amount of turning between the arms of the angle stated in terms of some unit.

In a numb er of text books the magnitude of ∡ **ABC** is written mag. ∡ **ABC**. However, in practice, we so often refer to the magnitude of angles, that for simplicity in these notes, **an ABC will mean the magnitude of that angle, unless otherwise stated.**

In order to eliminate any uncertainty as to which angle magnitude may be under consideration, it is good practice to indicate the turning that has taken place between the arms. This may be done using a curved line or arrow, as shown.

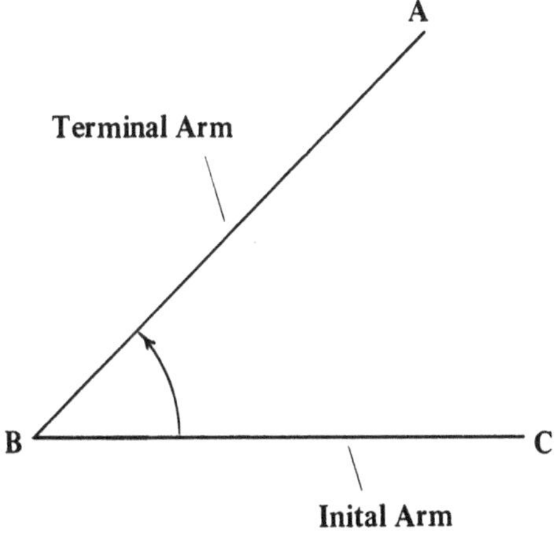

Note:

The arm of the angle that stays in the initial position, is called the **Initial Arm.**

Th are of the angle that rurns into a final position, is called the **Terminal Arm.**

`In everyday life we are constantly turning things or parts of things or causing things to turn.

Examples:

Screwing or unscrewing the cap of a toothpaste tube.
Dialing a telephone number.
Turning a key to unlock or close a locker.
Winding a watch.
Turning a door knob to open a door.
Turning a water tap on and off.
The wheel of your bike.

Write down at least 3 other thingsthat turn, rotate, revolve or spin.

1. ______________________________

2. ______________________________

3. ______________________________

Many articles have marks on them to help us to judge the amount of turn.

Examples:

The temperature control knob on a stove.

The face of a watch.

The channel selector on a television set.

Write down 3 other articles that have marks on them to help us to judge the amount of a turn.

1. ______________________________

2. ______________________________

3. ______________________________

It is frequently convenient to measure the amount of turning by taking one revolution as the unit. This is the mathematical name for one complete turn.

A revolution has a centre, a direction and a size of revolution.

You may have used an electric drill that has two speeds, 1,000 or 2,400 revolutions per minute (r.p.m.) This simply means that when the drill is being used, the drill spindle will make 1,000 or 2,400 complete turns per minute, depending on which speed has been selected.

In each of the following examples, wrtie down the number of complete turns made per minute.

1. An egg beater is revolving 650 revolutions per minute.

Ans: ______________

2. The circular blade of an electric saw is revolving at 7,800 r.p.m.

Ans: ______________

3. The propeller of an aircraft is making 11,700 r.p.m.

Ans: ______________

Practice Examples:

1. Consider the temperature control knob of an electric stove.

State the magnitude of the angle in revolutions and fractions oi a revolution when the pointer is moved from

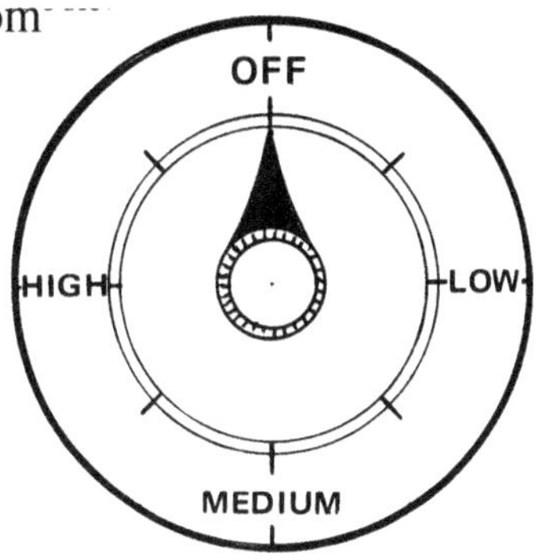

OFF
HIGH
LOW
MEDIUM

(a) the "Off" to the "Med" position ( )

(b) the "Med" to the "Low" position ( )

(c) the "High" to the "Off" position. ( )

(d) midway between the "Med" & "Low" position to the "Off" position. ( )

2. Imagine that you have a pocket compass and are standing facing Nprth. (Use the Mathomat compass points Fig. 23).

State the magnitude of the angle in revolutions or fractions of revolution when you turn:

Answers

(a) in an anticlockwise direction to face W ( )

(a) in an anticlockwise direction to face SW ( )

(a) in an anticlockwise direction to face SSE ( )

(a) in an anticlockwise direction to face NNE ( )

Answers

(a) in a clockwise direction to face NNE ( )

(a) in a clockwise direction to face ESE ( )

(a) in a clockwise direction to face WSW ( )

3. What direction from North would you be facing if you turned:

(a) $1/4$ of a revolution in an anticlockwise direction ( )

(b) $3/4$ of a revolution in an anticlockwise direction ( )

(c) $1/2$ of a revolution in an anticlockwise direction ( )

(d) $1/16$ of a revolution in an anticlockwise direction ( )

(e) $5/16$ of a revolution in an anticlockwise direction ( )

(f) $3/16$ of a revolution in a clockwise direction ( )

(g) $5/8$ of a revolution in a clockwise direction ( )

(h) $1/16$ of a revolution in a clockwise direction ( )

4. Consider now a clock face. **7**

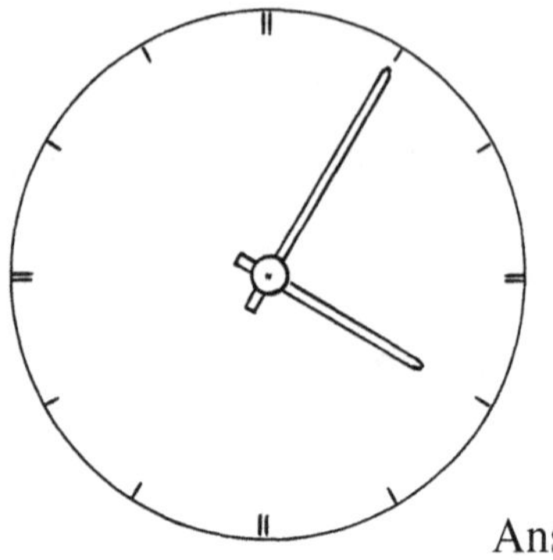

State the magnitude of the angle, in revolutions or fracttions of a revolution, the minute hand of the clock turns in:

	Answers		Answers
(a) 30 minutes	( )	(b) 45 minutes	( )
()c) 1 hour	( )	(d) 5 minutes	( )
(e) 20 minutes	( )	(f) 35 minutes	( )
(g) 2 hours	( )	(h) 16 minutes	( )
(i) 10 minutes	( )	(j) 3 hours 15 minutes	( )

5. A sewing machine makes 2 stitches for ever turn of the flywheel. State the magnitude of the angle in revolutions or fractions of revolution which thee flywheel turns through in making:

	Answers		Answers
(a) 14 stitches	( )	(b) 3 stitches	( )
9c) 7 stitches	( )	(d) 45 stitches	( )
(e) 1 stitches	( )	(f) 35 stitches	( )

6. A ferris wheel at an amusement park revolving at 6 revolutions per minute. State the magnitude of the angle turned through in:

	Answers		Answers
(a) 3 minutes	( )	(b) 30 seconds	( )
()c) 10 seconds	( )	(d) 15 seconds	( )
(e) 5 seconds	( )	(f) 2.5 seconds	( )

RIGHT ANGLES

When an interval is rotated through one quarter of a revolution the angle formed is called a **Right Angle**

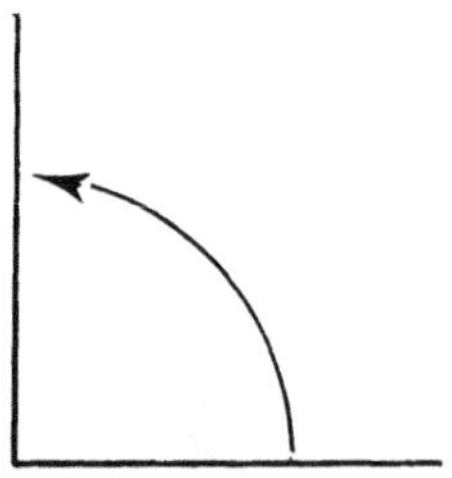

This angle is constantly in use in building, engineering and drafting.

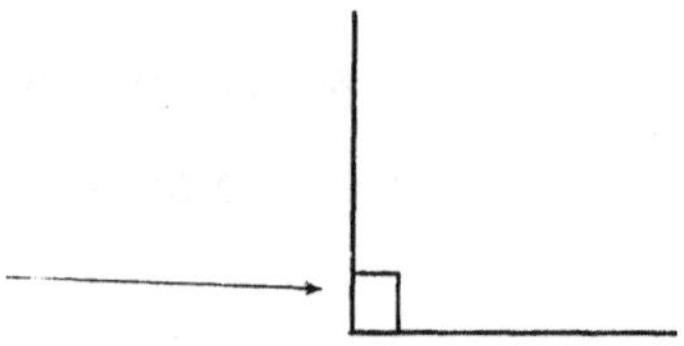

To indicate that an angle is a Right Angle, it is often marked with a small square.

The following are Right Angles: mark each accordingly.

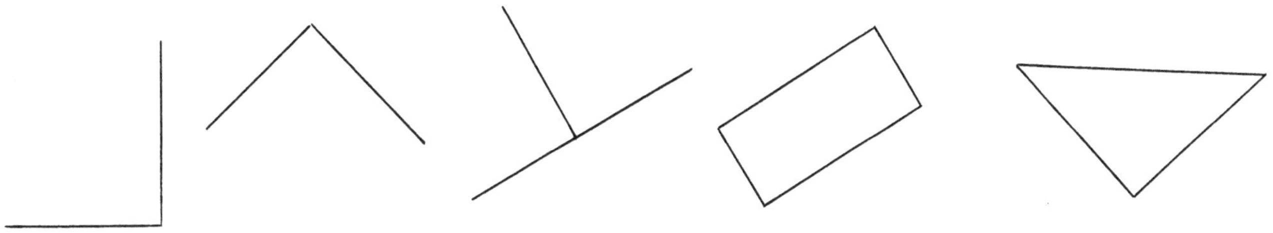

An accurate right-angle can be made from a piece of paper without measuring the rotation. Try this – take a piece of plain paper roughly circular and approximately 10cm across. Fold, as shown, so that X meets Y.

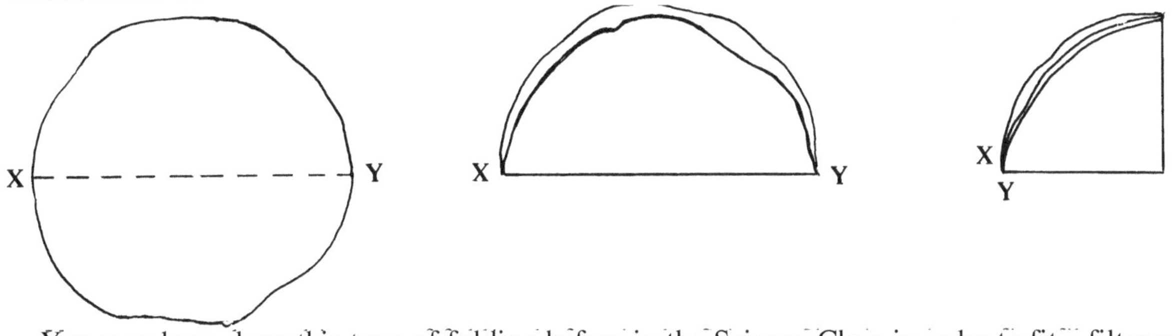

X Y X Y X Y

You may have done this type of folding before in the Science Class in order to fit a filter paper into a funnel.

Use this right angle to check the corners of books, desk tops, set squares, rulers, floorboard joins, etc.

Now unfold the piece of paper. You have divided a complete revolution into quarters, and the crease lines have formed four right angles. Carefully mark the crease lines and right angles with a pencil or biro, and attach this piece of paper to this sheet or to your book.

Lines are said to be Perpendicular to each other if they meet or intersect at right angles.

Which of these pairs of lines are perpendicular to each other?

A B C D E F

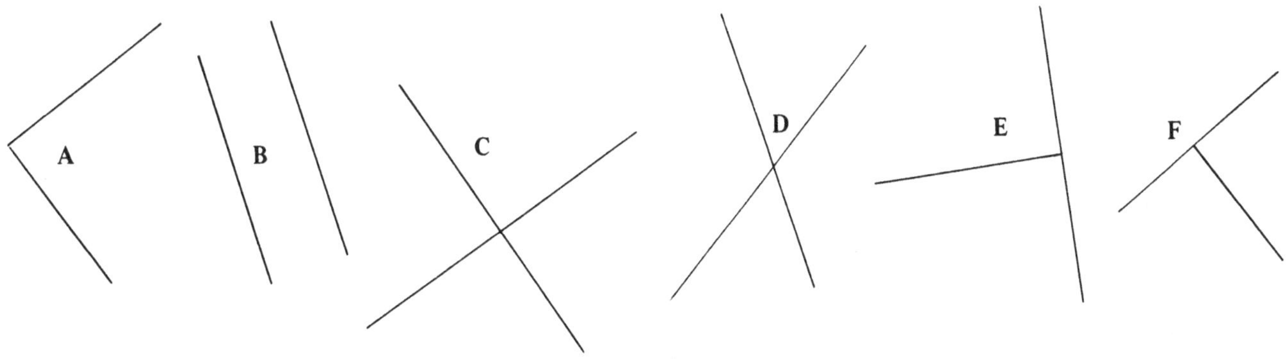

Answer:

Types of Angles:

The magnitude of an angle determines the name given to that angle

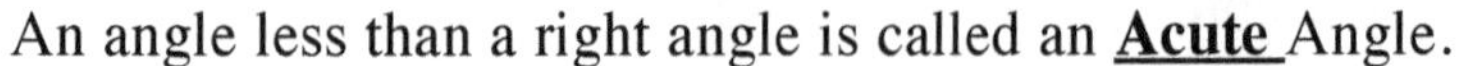
An angle less than a right angle is called an **Acute** Angle. 9

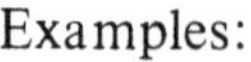
Examples:

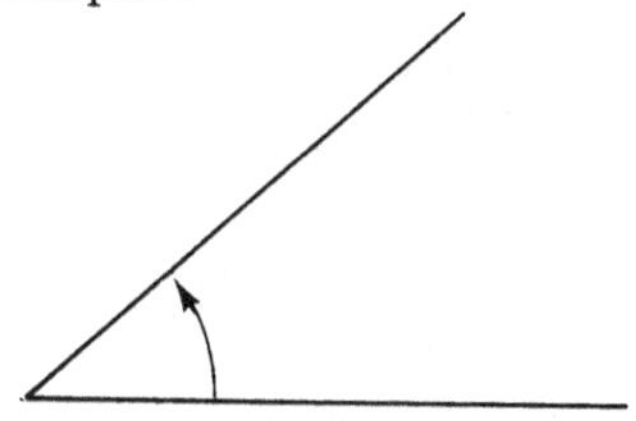
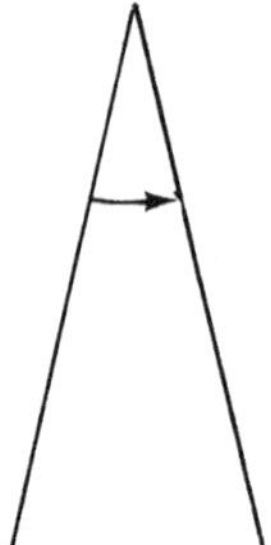
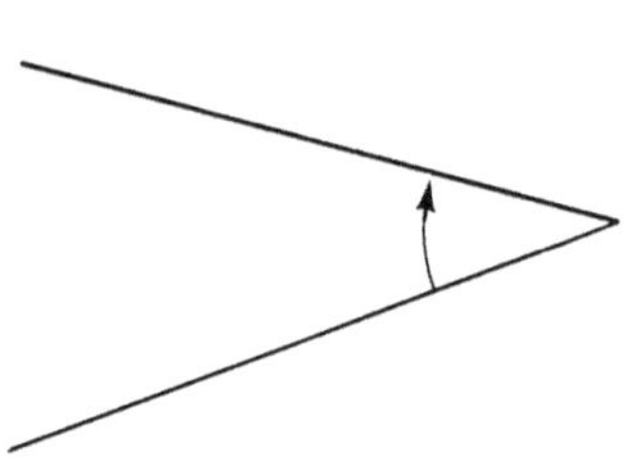

An angle greater than one right angle but less than two right angles is called an **Obtuse** Angle.

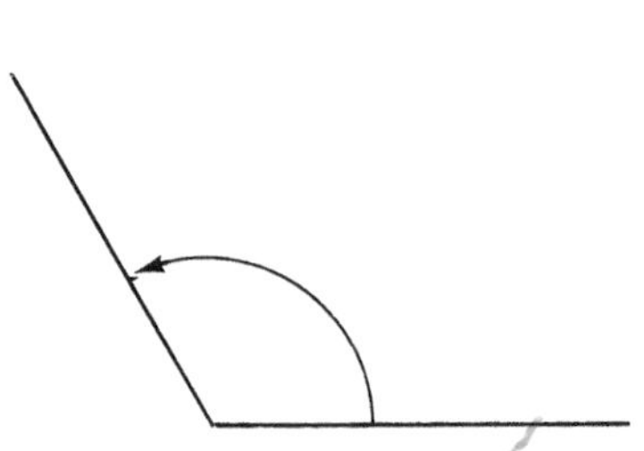
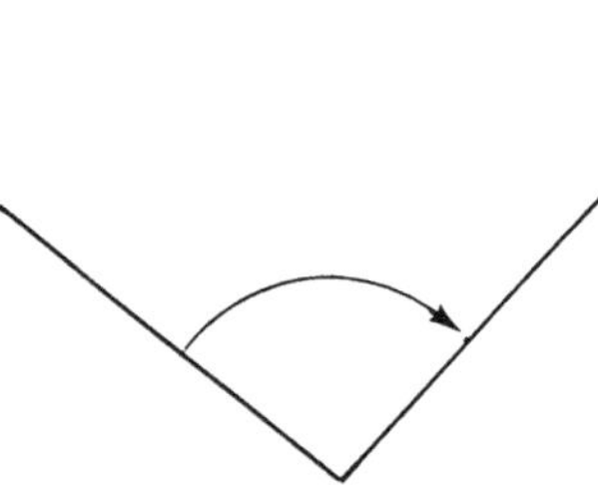
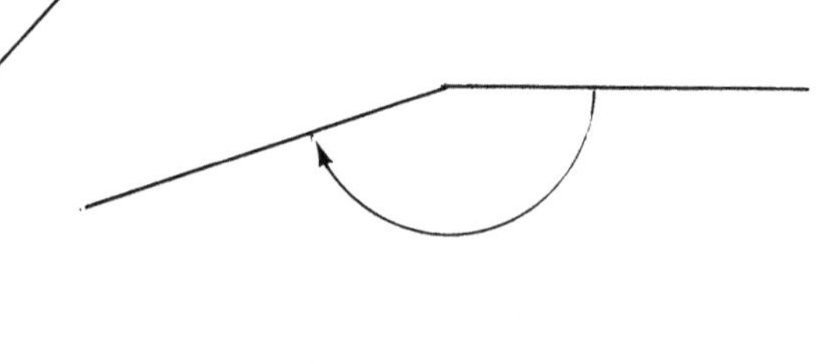

An angle that is exactly equal to two right angles is called a **Straight** Angle.

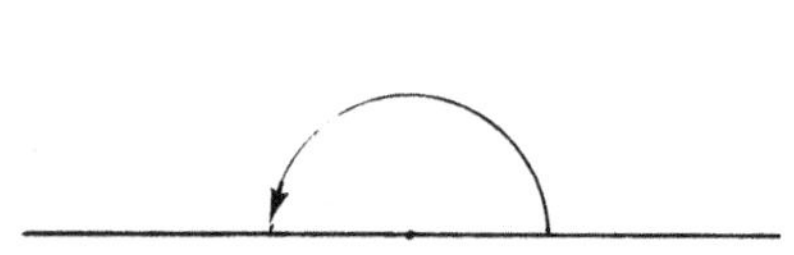
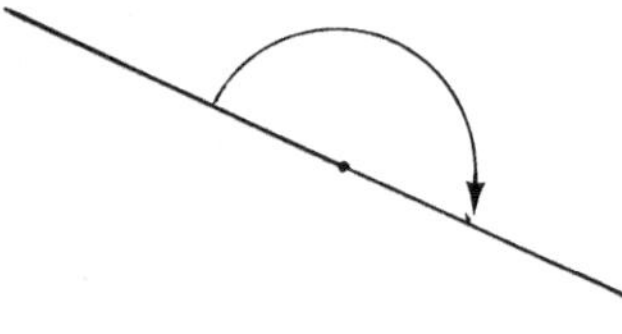
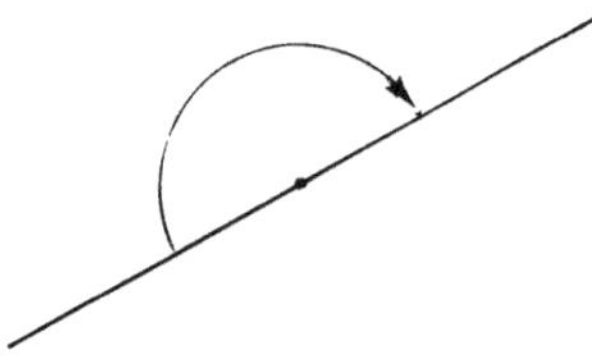

An angle that is greater than two right angles but less than four right angles is called a **Reflex** angle.

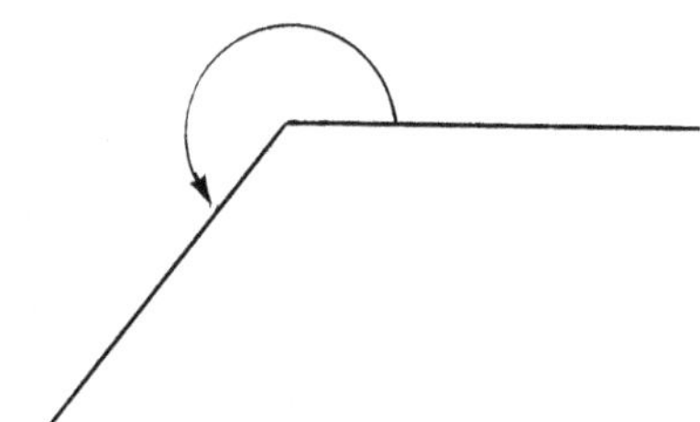
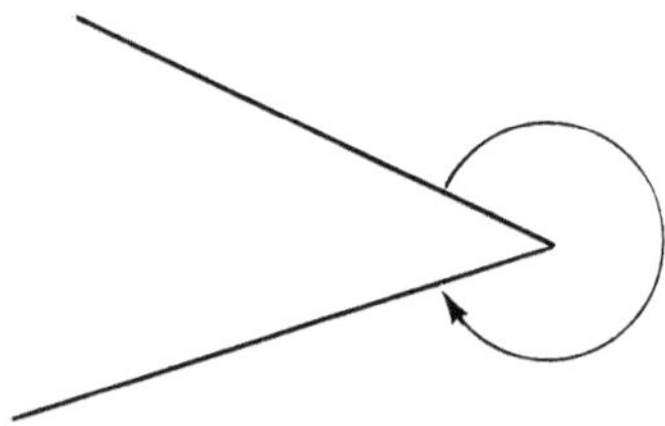
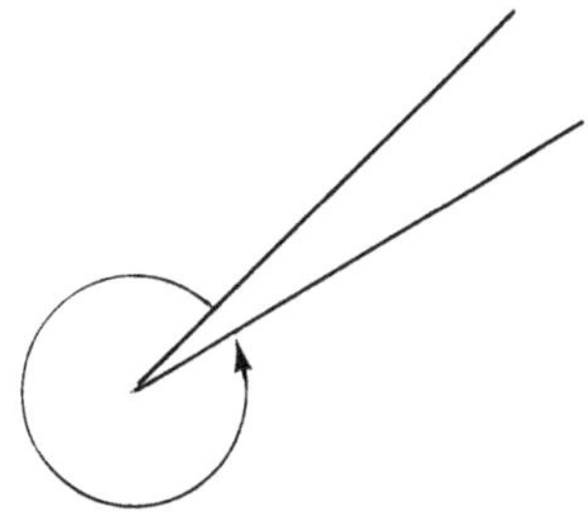

An angle that is exactly equal to four right angles is called a **Revolution.**

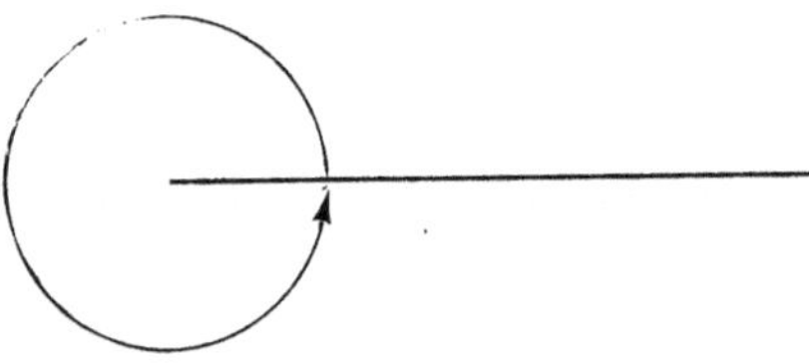

In each of the above angles, name (in red biro) both the Initial Arm and the Terminal Arm.

Practice Examples: **10**

1. In the space provided, draw a freehand sketch of an angle of magnitude. (In each case, the initial arm has been drawn).

(a) 1 right angle	(b) 3 right angles	(c) $\frac{1}{2}$ of a right ∡	(d) $\frac{1}{3}$ of a right ∡
(e) $\frac{1}{6}$ of a right ∡	(f) $1\frac{2}{3}$ right ∡	(g) $2\frac{1}{3}$ right ∡	(h) $\frac{5}{6}$ of a right ∡

2. Complete the table on the next page by selecting which of the following angles are acute, right angled, obtuse, straight or reflex.

a. 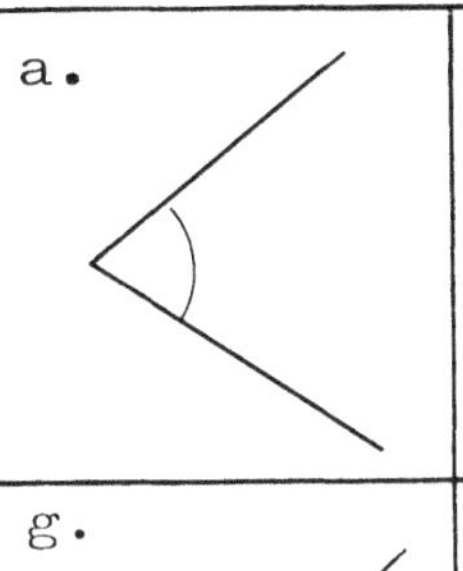	b. 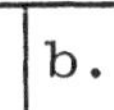 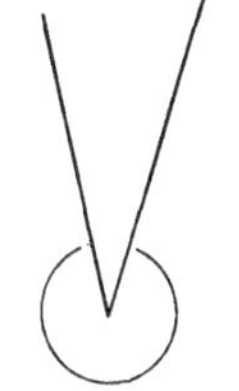	c. 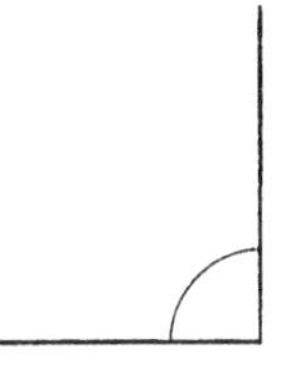	d. 	e. 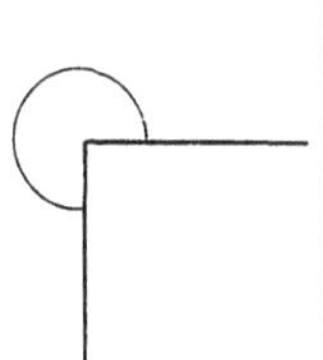	f. 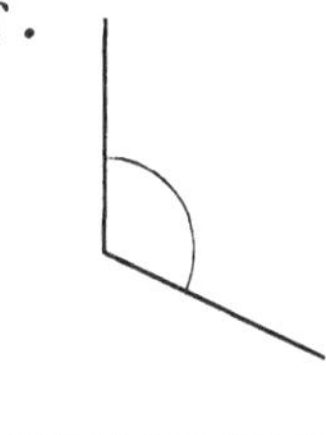
g. 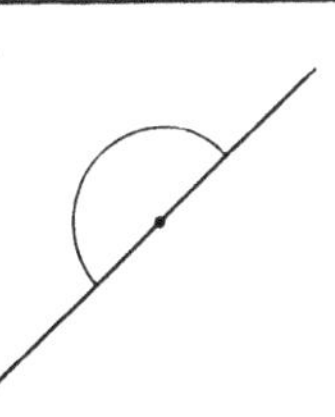	h. 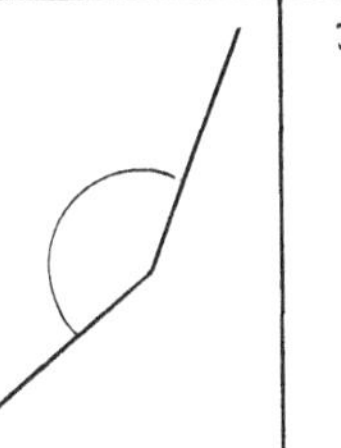	i. 	j. 	k. 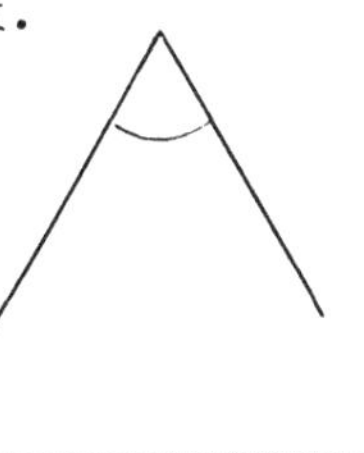	l. 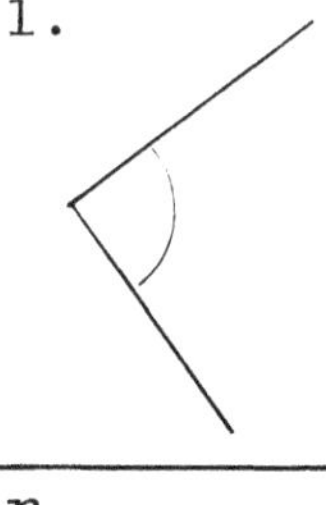
m. 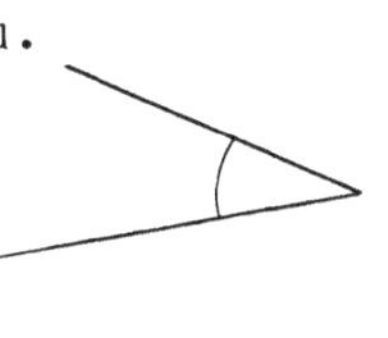	n. 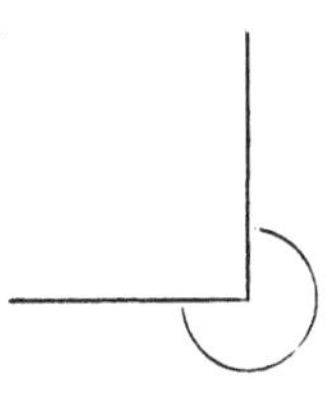	o. 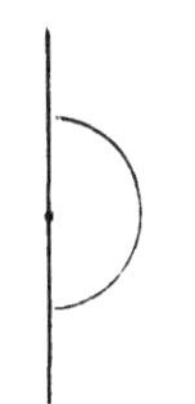	p. 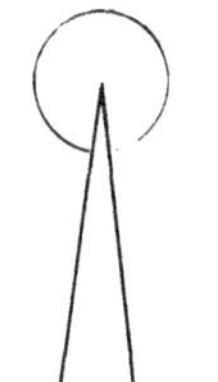	q. 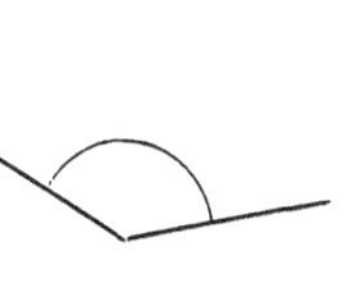	r. 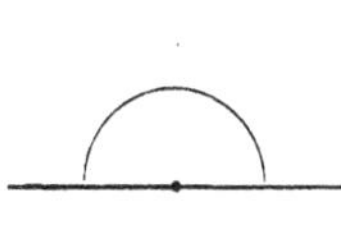
s. 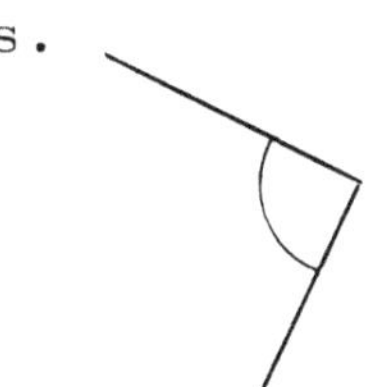	t. 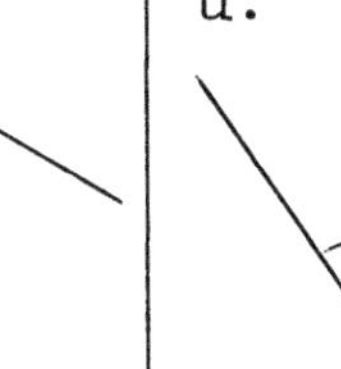	u.	v. 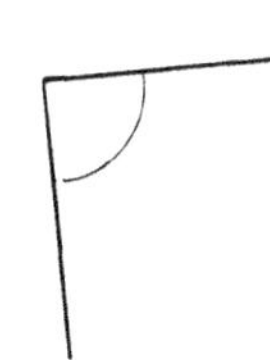	w. 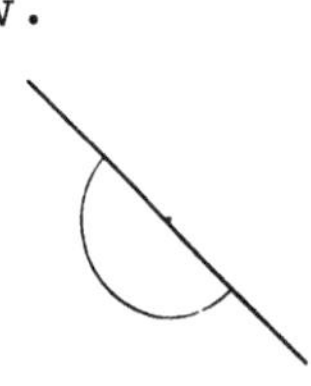	x. 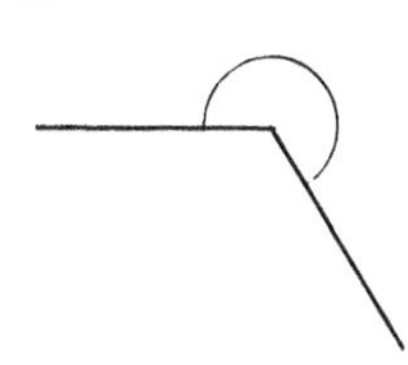

ANGLES

Acute	Right	Obtuse	Straight	Reflex
a.				b.

DEGREE MEASURE

Until now, we have only used one revolution as the unit of measure for the magnitude of an angle. In practice, surveyors, engineers, navigators, astronauts, require greater accuracy. and hence they find it more convenient to use a smaller unit called the ***Degree.***

Here is an angle of 1 degree.

A Degree is $\frac{1}{360}$ of one revolution.

i.e., there are 360 degrees in one revolution.

It is interesting to note at this stage, that this unit probably owes its origin to the early Babylonian astronomers (3000 to 2000 years BC) who worked with accuracy to measure and describe the movements of the planets. They were very interested in sixties and thus probably took one-sixth of a turn and divided it into sixtieths.

Sixty degrees is one sixth of one revolution.

$\frac{1}{6}$ of 1 Rev.

A short way of writing degrees:

45 degrees is written 45°   60 degrees is written 60°

Change to degrees and write in the short form

(a) 1 right ∡ Ans. ________ (b) 3 right ∡ Ans. ________

(c) $\frac{1}{2}$ right ∡ Ans. ________ (d) 4 right ∡ Ans. ________

(e) $\frac{2}{3}$ right ∡ Ans. ________ (f) $2\frac{1}{3}$ right ∡ Ans. ________

(g) $3\frac{1}{6}$ right ∡ Ans. ________ (h) $1\frac{3}{5}$ right ∡ Ans. ________

(i) $4\frac{4}{9}$ right ∡ Ans. ________ (j) $\frac{5}{6}$ right ∡ Ans. ________

The following represents a degree scale (marked in units of 30•) similar to a clock face without the hands.

Complete both the inner (anticlockwise) scale, and the outer (clockwise) scale.

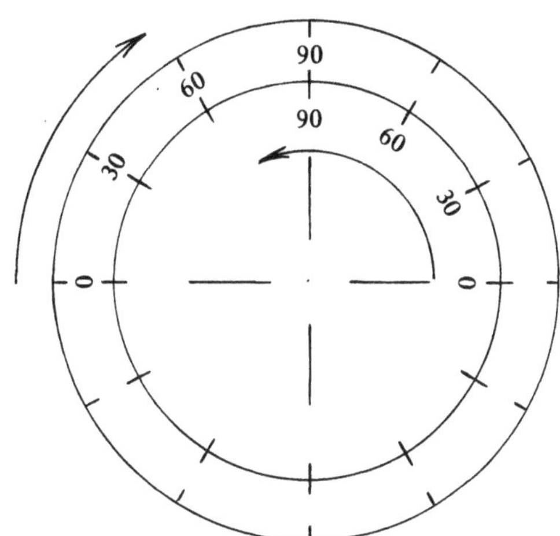

Now complete the following table:

Measure of angle in Revolutions	1	$\frac{1}{2}$		$\frac{1}{8}$	$\frac{3}{4}$		$\frac{1}{6}$		$\frac{1}{24}$
Measure of angle in Degrees	360		90			120		30	

Estimating

How good are you now at estimating the magnitude of an angle?

Without using your Mathomat protractor, try to estimate as accurately as you can the magnirude of the following angles. Write your answer within the angle, as shown.

Perhaps your folded paper right angle or the markings on the face of your wrist watch, may help.

Examples:

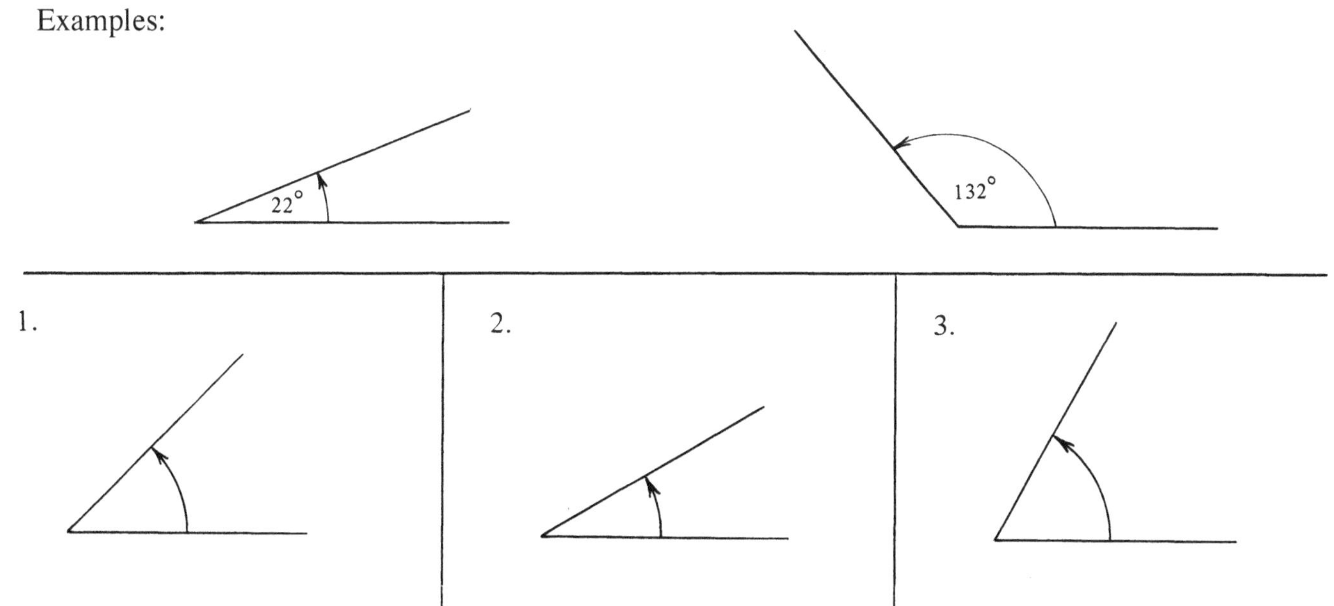

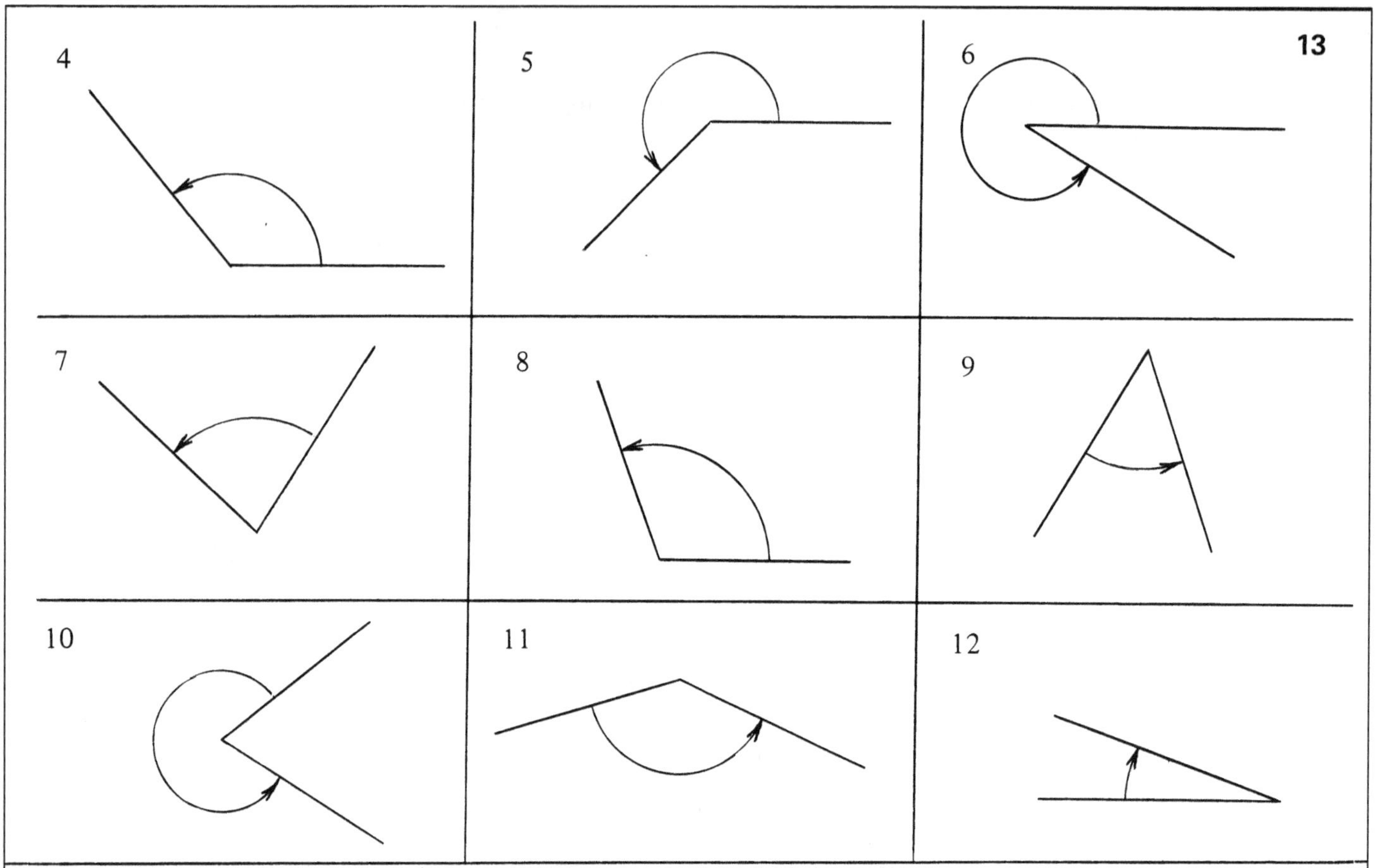

13
4
5
6
7
8
9
10
11
12

Note: A lower case letter of the alphabet is often used to indicate the magnitude of an unknown angle.

Examples:

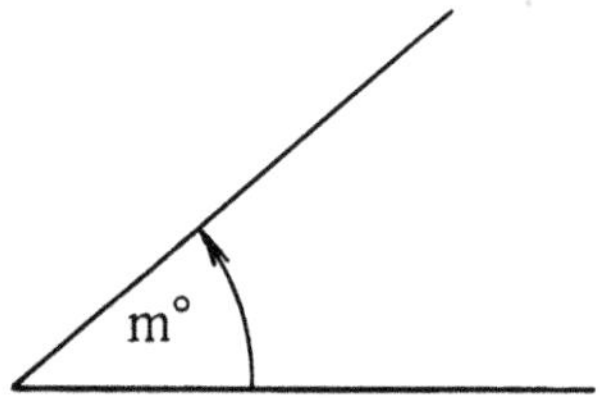

m°

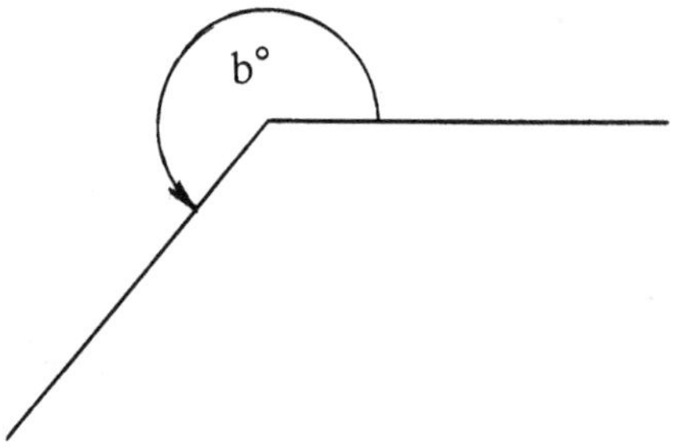

b°

Estimate the magnitude of the following marked angles.

a°

Ans. a° = °

y°

Ans. y° = °

w°

Ans. w° = °

k°

Ans. k° = °

h°

Ans. h° = °

n°

Ans. n° = °

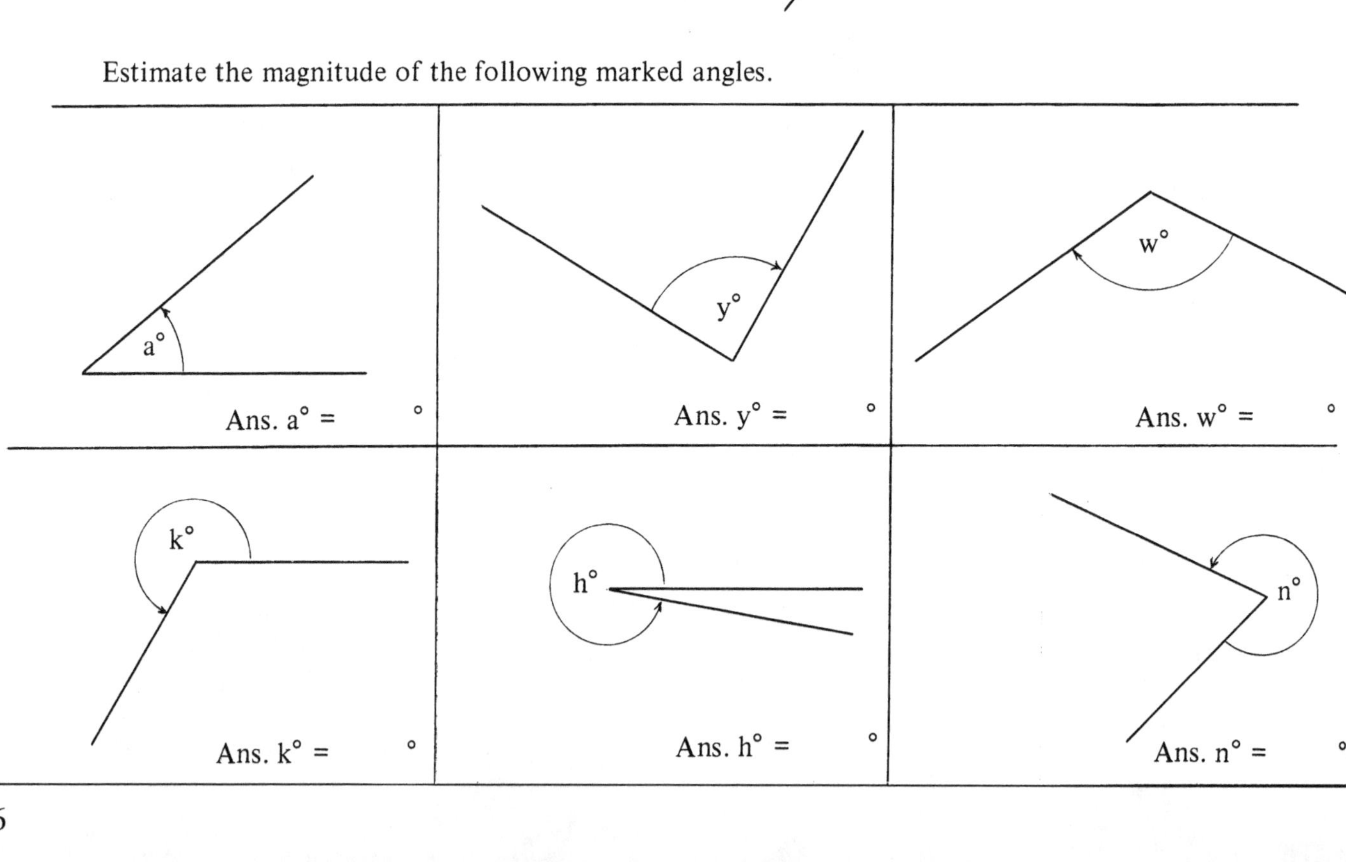

Use of MATHOMAT PROTRACTOR

Making an accurate guess as to the magnitude of an angle (ie: estimating) is an important part of measuring.

The Mathomat protractor is necessary for precise measurement of angles, however it is important to keep in the habit of estimating angle sizes while you are working with a protractor.

Now examine very carefully the protractor located in the centre of your Mathomat and note the following features:

The protractor has a central locating hole (to find the centre of a circle or the vertex of an angle), a centre or zero line, an outer degree scale (clockwise) and an inner bearing scale (anti-clockwise).

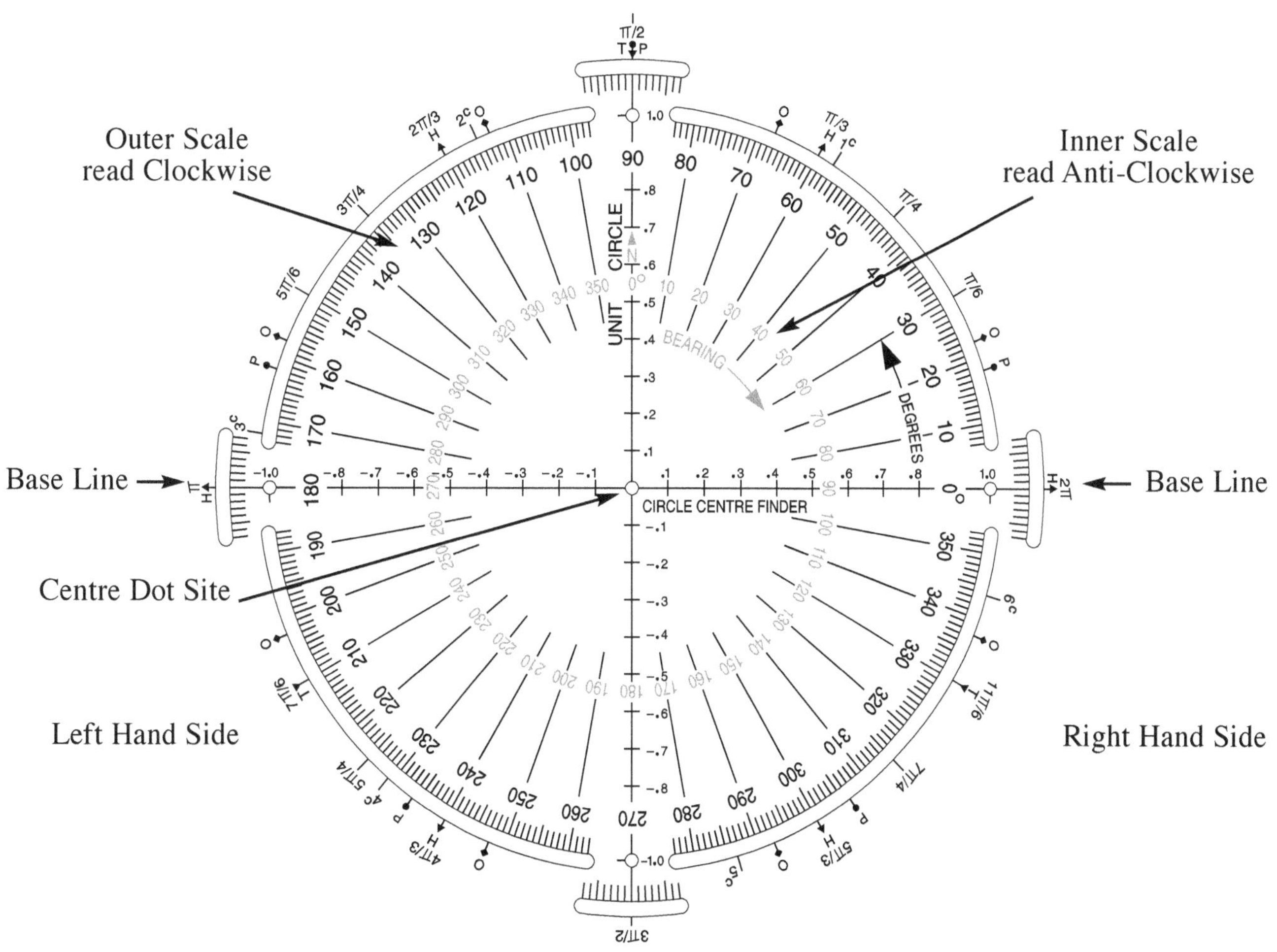

To be able to measure the magnitude of an angle ABC it is important to remember the following terms:

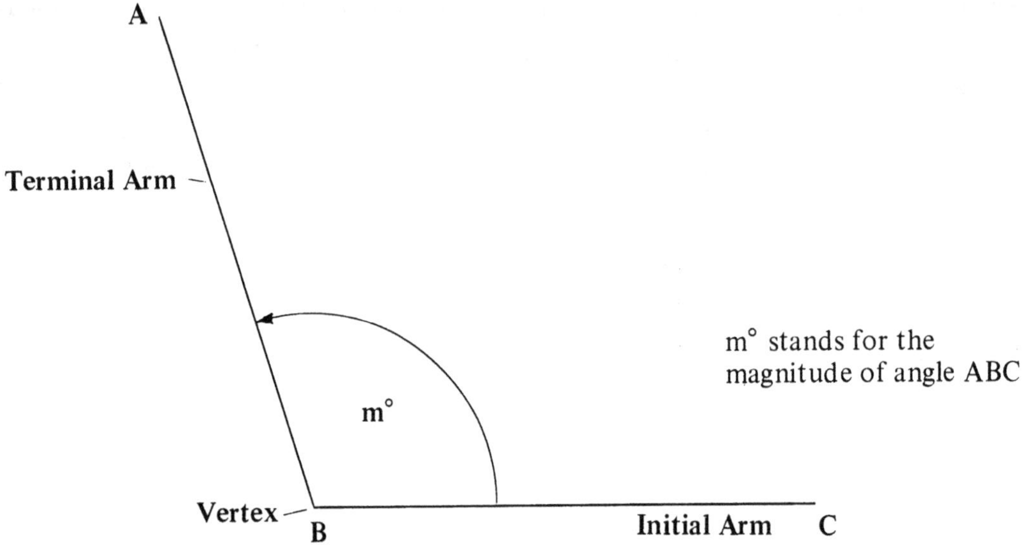

The main idea is to find out what angle on the protractor corresponds to the angle marked on the paper.

Note: Estimate the angle first before using your protractor.

The Flow Diagram on the following page gives detailed steps of how to measure the magnitude of an angle < 180° with your protractor.

To Measure the Magnitude of any Angle, ⩽ 180°, with a Protractor.

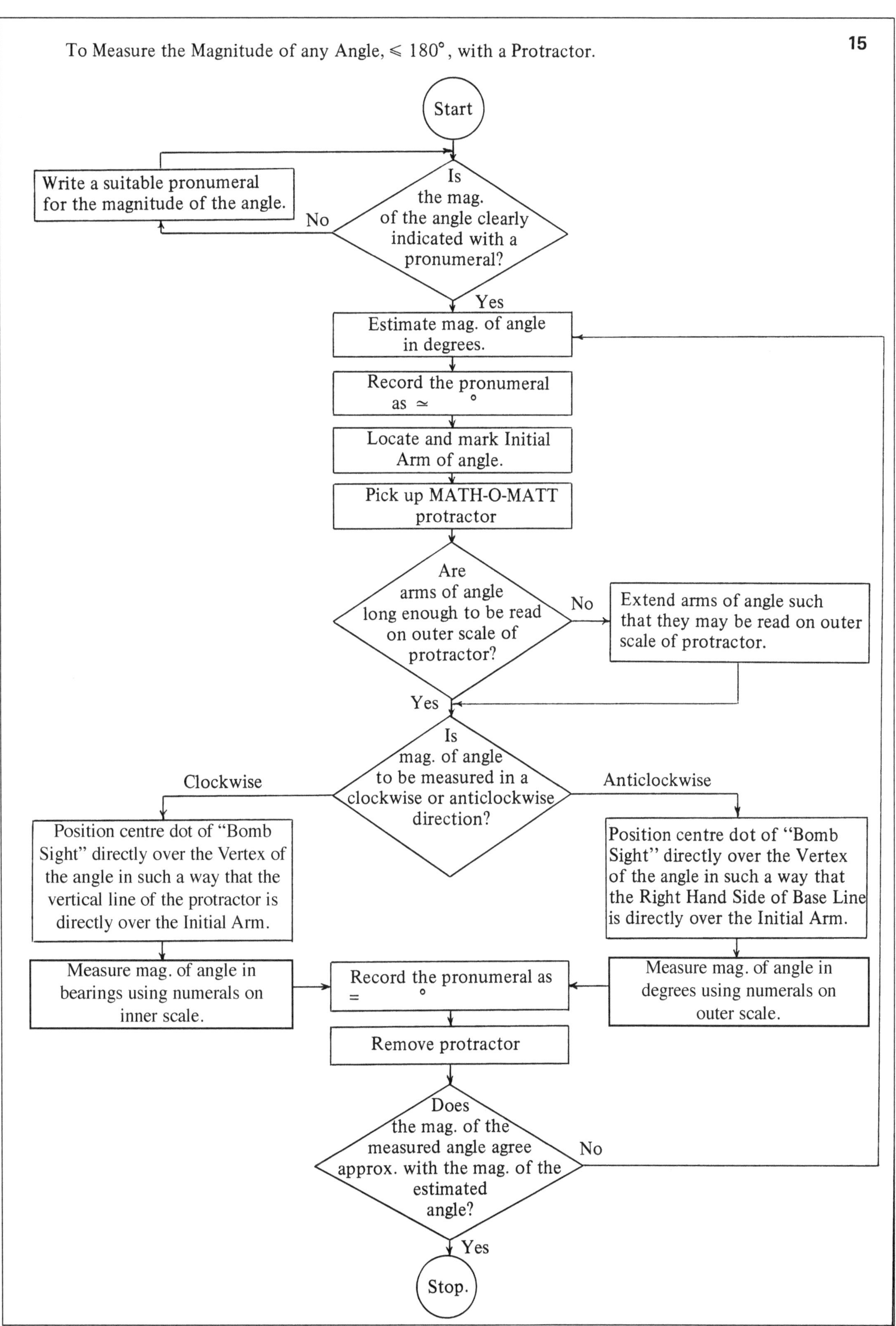

16

In each case, use your Mathomat protractor to measure the magnitude of the angle indicated with a pronumeral. Write your answer in a similar manner to the worked example shown.

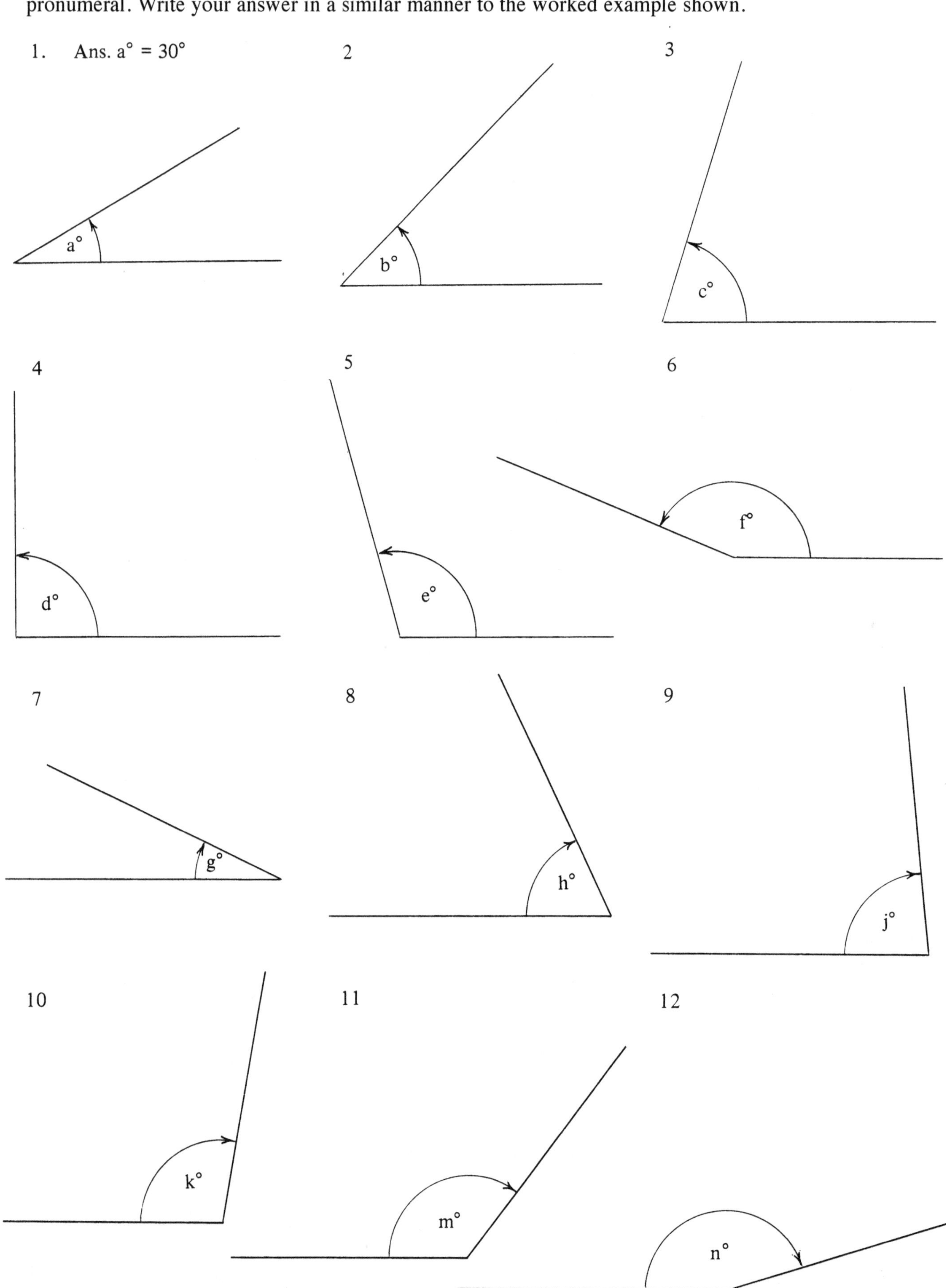

Further Examples:

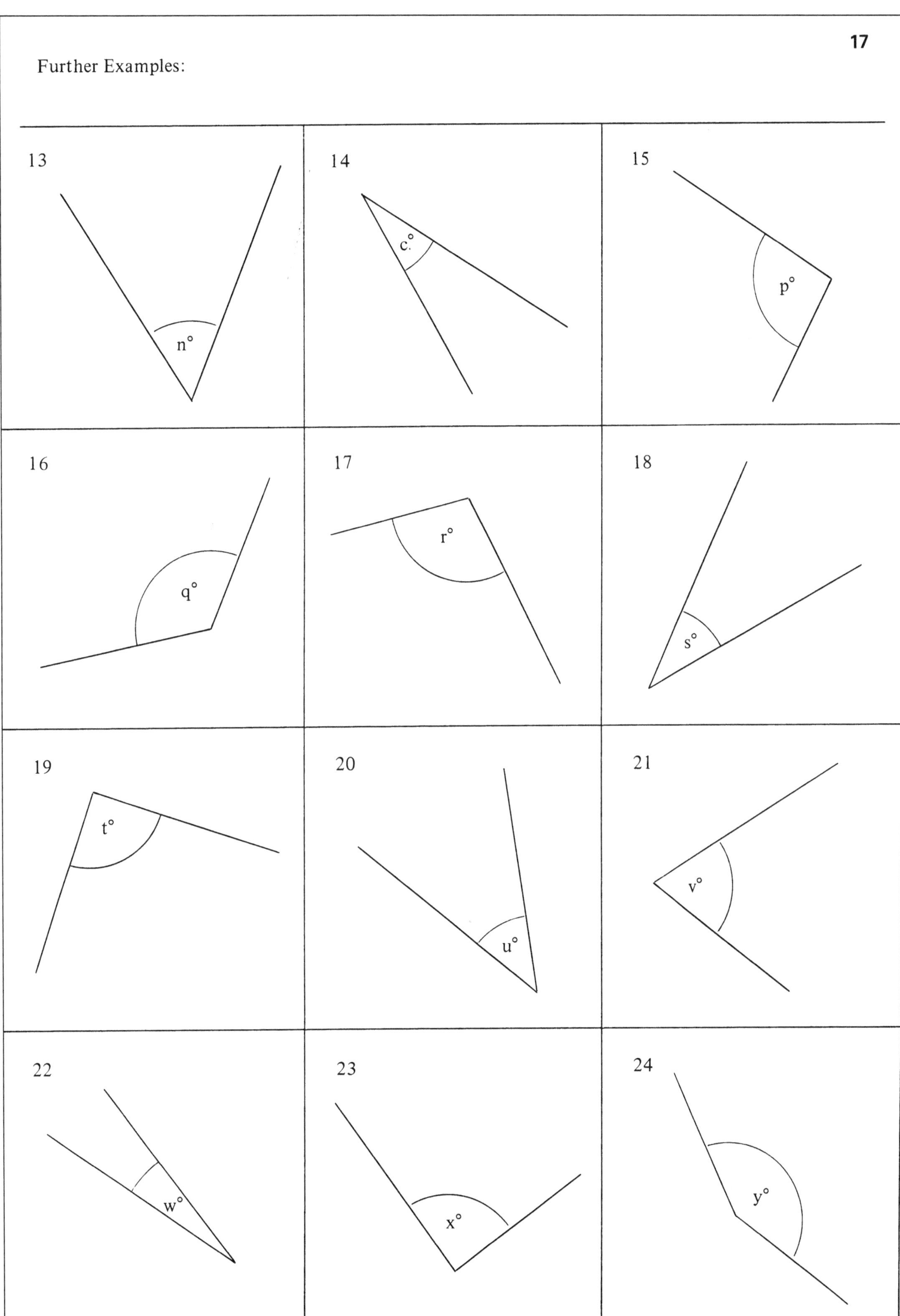

Problem: How would you measure a reflex angle?

Answer: One method is to simply measure the magnitude of the angle that is less than 180° and then subtract this angle from 360°.

Example: Measure the magnitude of the reflex angle ABC

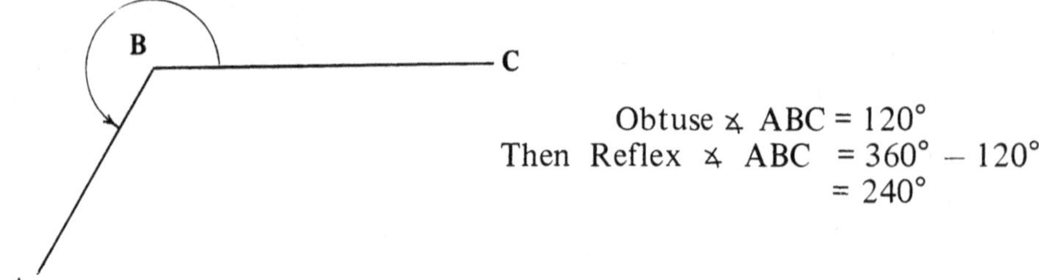

Obtuse ∢ ABC = 120°
Then Reflex ∢ ABC = 360° – 120°
= 240°

Use your Mathomat protractor to measure the magnitude of the folling reflex angles.

1 (D, C, E) Ans. °	2 (F, G, H) Ans. °	3 (J, K, L) Ans. °
4 (M, N, O) Ans. °	5 (P, Q, R) Ans. °	6 (U, T, S) Ans. °
7 (V, W, X) Ans. °	8 (A, B, Y) Ans. °	9 (G, H, K) Ans. °

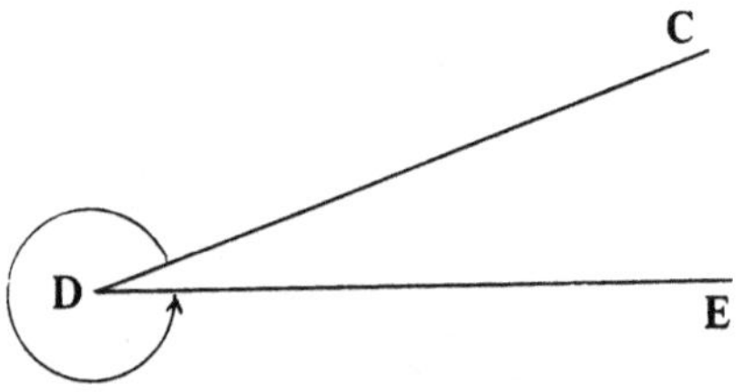

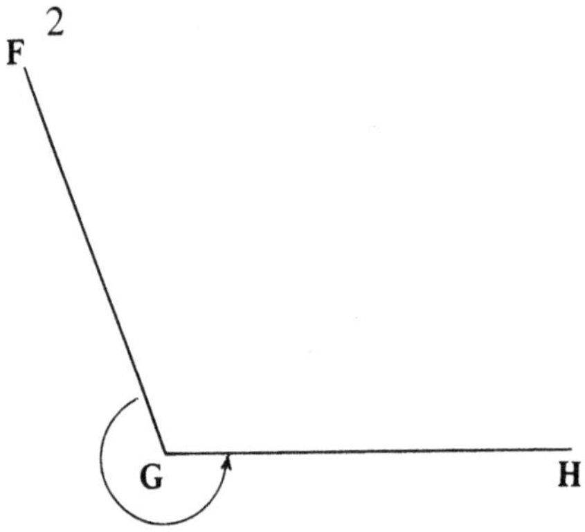

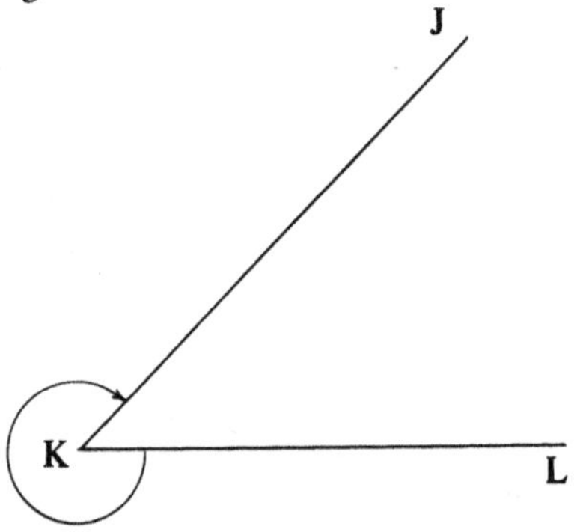

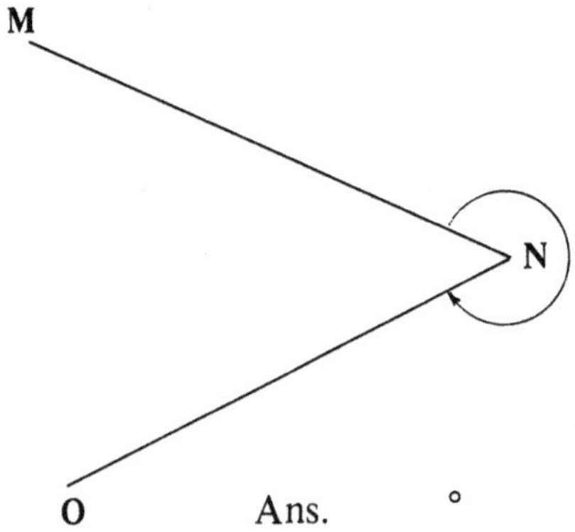

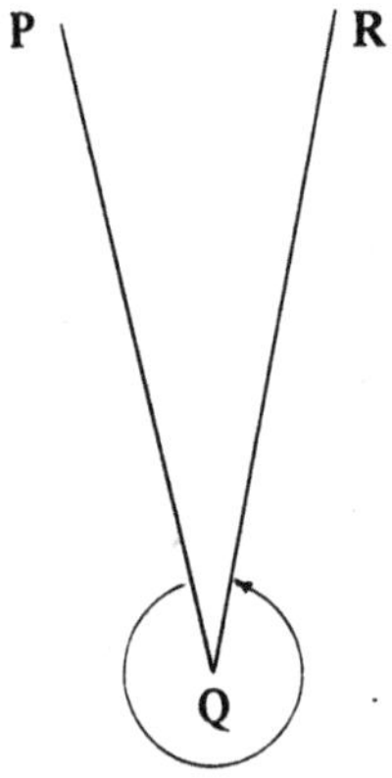

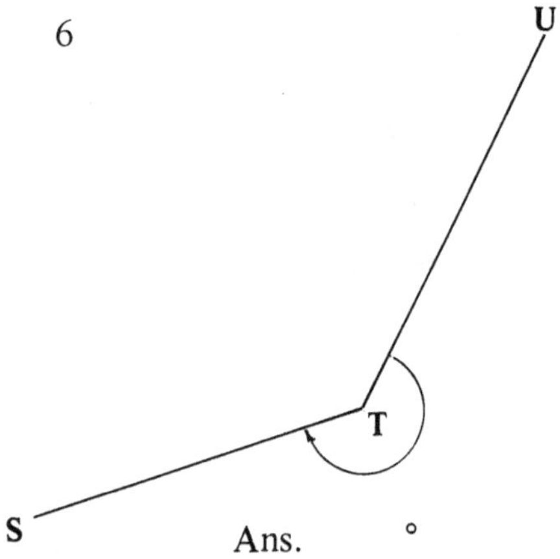

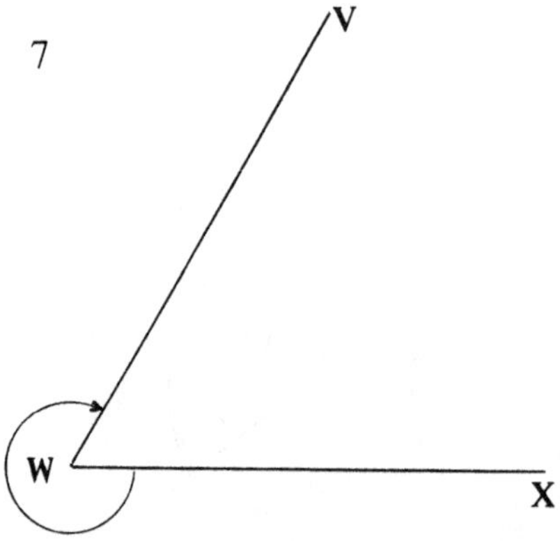

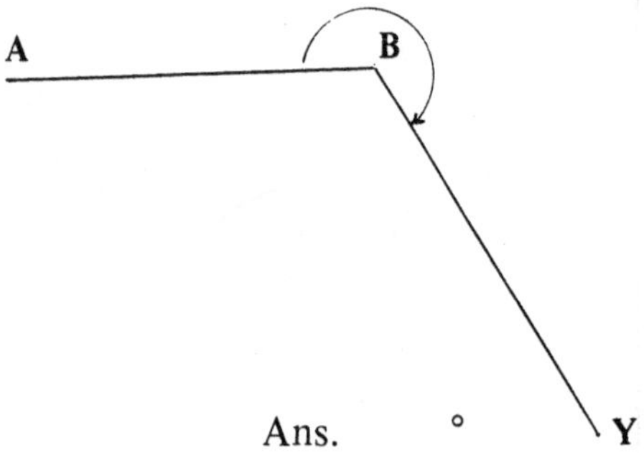

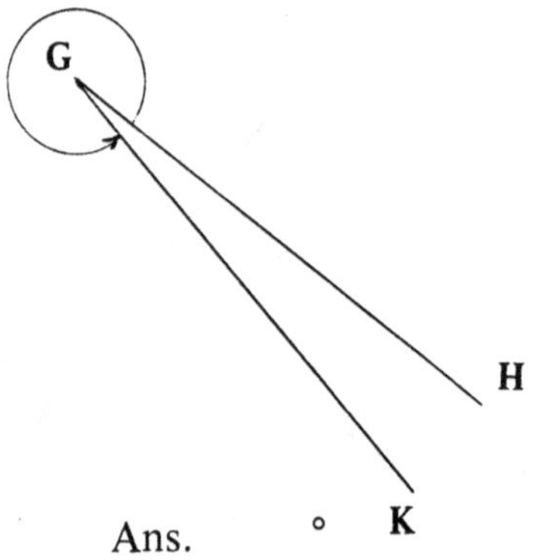

19

Using the MATH-O-MATT Protractor to draw any angle ⩽ 180°.

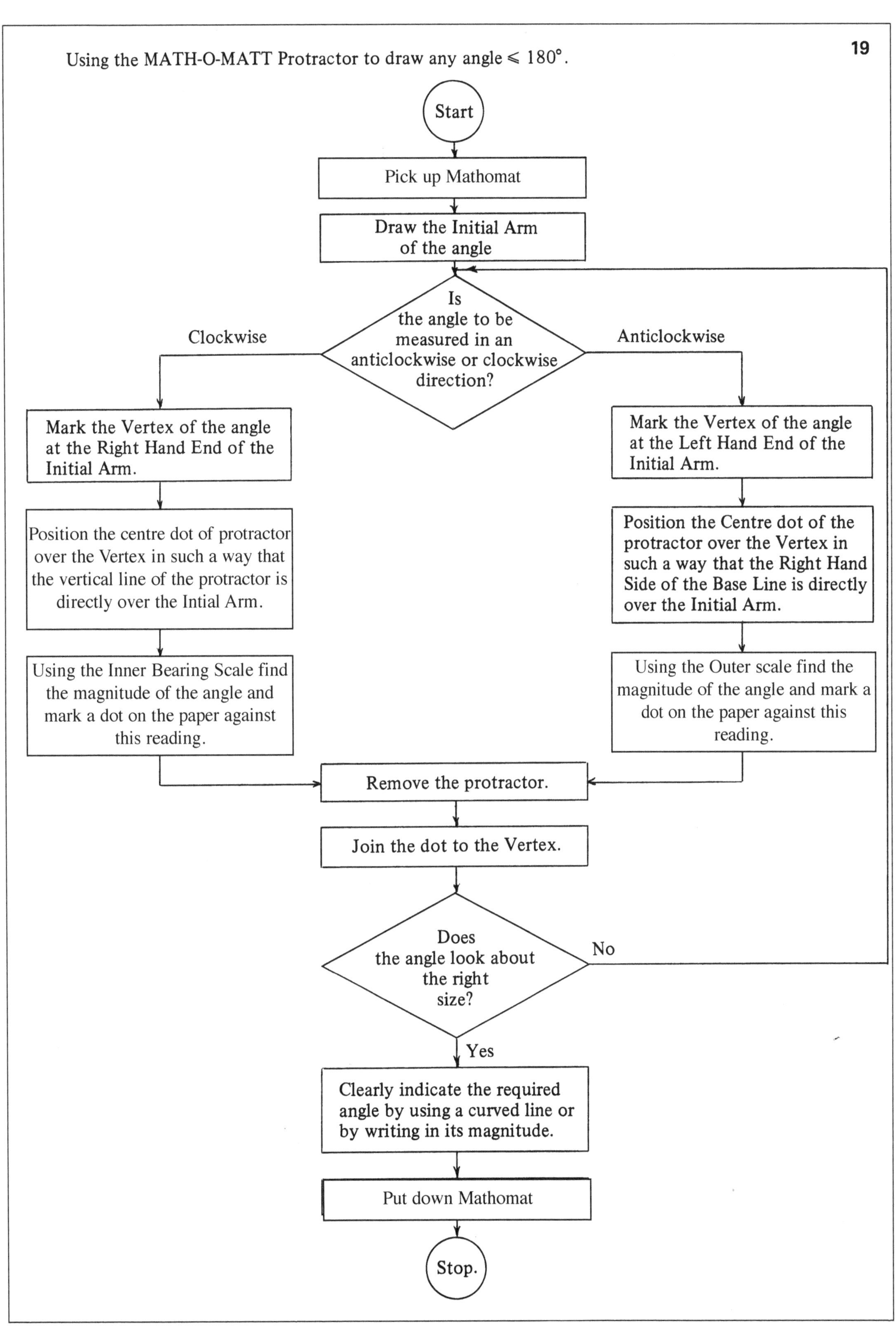

20

Practice Examples:

Use your Mathomat to draw and mark angles of the following magnitudes in the space provided.

Example: 48° measured anticlockwise

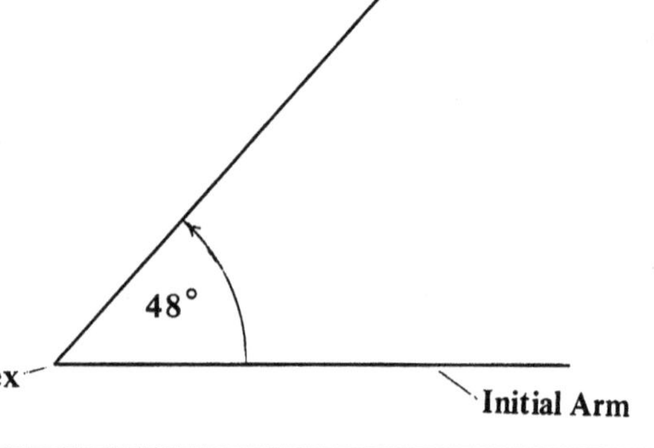

A 25° measured clockwise

B 57° measured anticlockwise

C 90° measured clockwise

D 112° measured anticlockwise

E 134° measured clockwise

F 167° measured anticlockwise

Problem: How would you draw a reflex angle?

Answer: One method is to simply subtract the magnitude of the relfex angle from 360°, then draw the resulting acute or obtuse angle, clearly marking in the reflex angle.

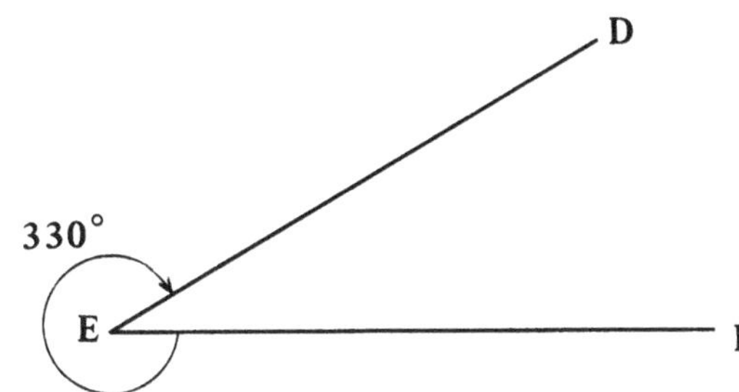

Example:

Draw a reflex angle DEF of 330°
Now the Acute ∢DEF = 360° − 330°
= 30°

Draw this angle and then mark the reflex angle.

Use your Mathomat protractor to draw the following reflex angles in the space provided.

A 291° **B** 227°

C 314° **D** 353°

E 197° **F** 277°

ADJACENT ANGLES

Two angles that share a common arm and have a common vertex are called Adjacent Angles. (That is, they are side by side).

Example:

A
B C
D

Angles ABC and CBD are Adjacent to each other

In each case write down two angles that are adjacent to angle WXY

1

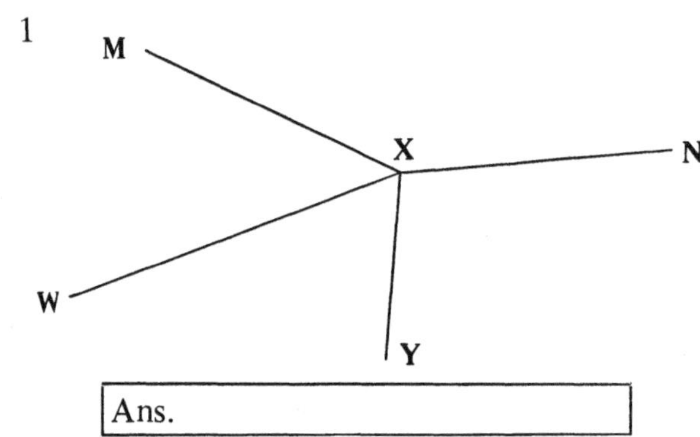

2

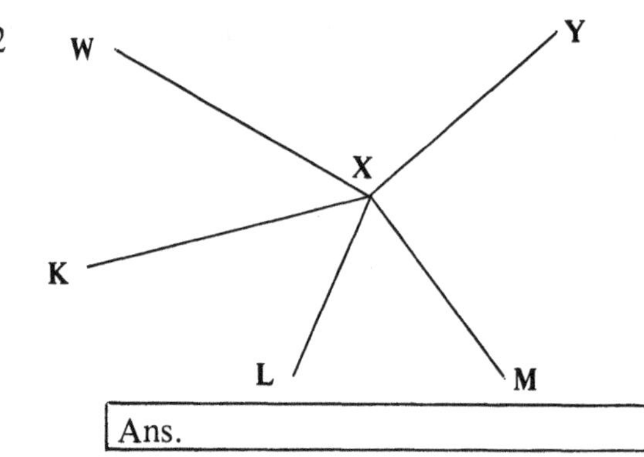

COMPLEMENTARY ANGLES

In each case calculate the size of the angles marked with a pronumeral. Do not use your protractor.

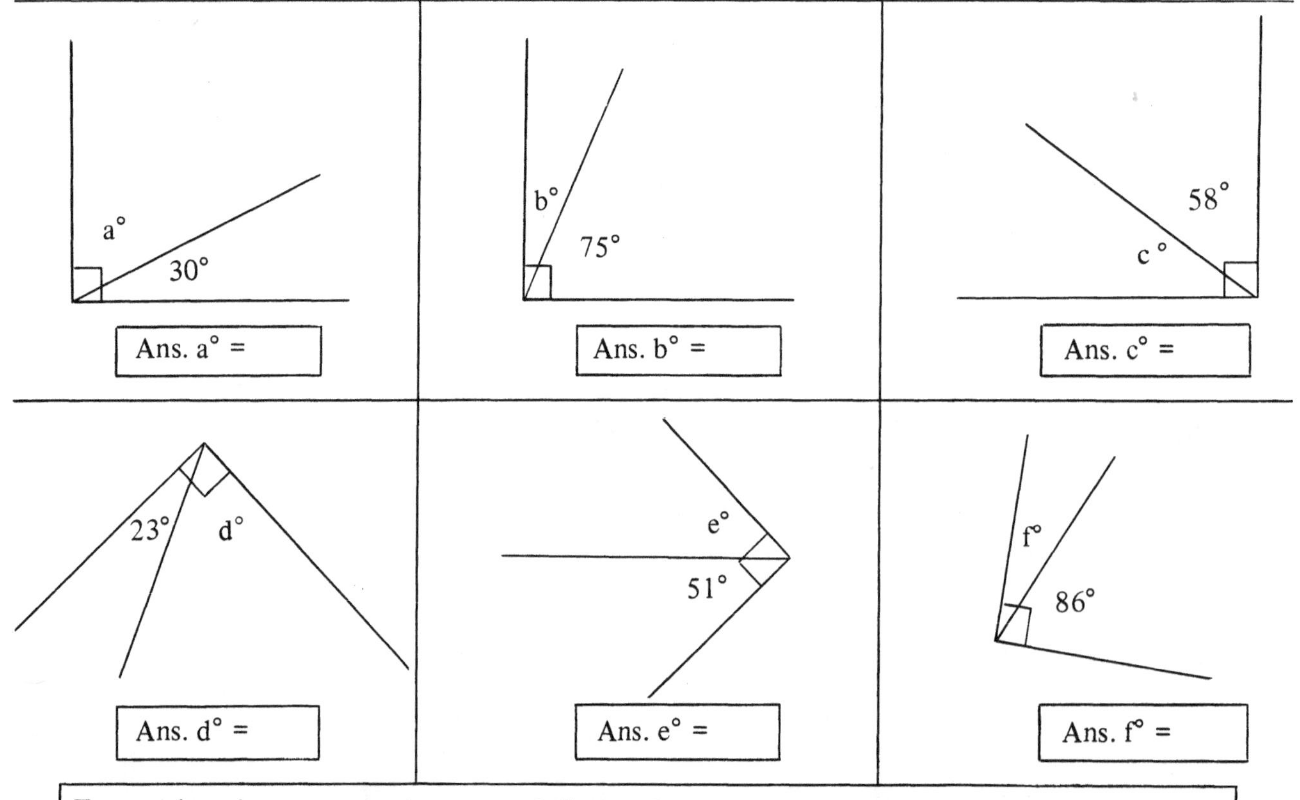

Two angles whose magnitudes sum to 90° (One Right Angle) are called Complementary Angles.

e.g. 30° is the complement of 60° and 60° is the complement of 30°.

Practice Examples:

1. In each caSe, write down the complement of the angle stated.

(a) 72° Ans. ( ) (b) 1° Ans ( ) (c) 45° Ans ( )

(d) 37° Ans. ( ) (e) 83° Ans. ( ) (f) 69° Ans ( )

2. State in the box provided, whether the following statements are true or false.

(a) 37° and 53° are complemen tary ☐

(b) 80° and 9° are complementary ☐

(c) 47° and 45° are complementary ☐

(d) 19° is the complement of 81° ☐

(e) 78° is the complement of 12° ☐

(f) 53° is the complement of 37° ☐

SUPPLEMENTARY ANGLES

In each case, calculate the size of the angles marked with a pronumeral. Do not use your protractor.

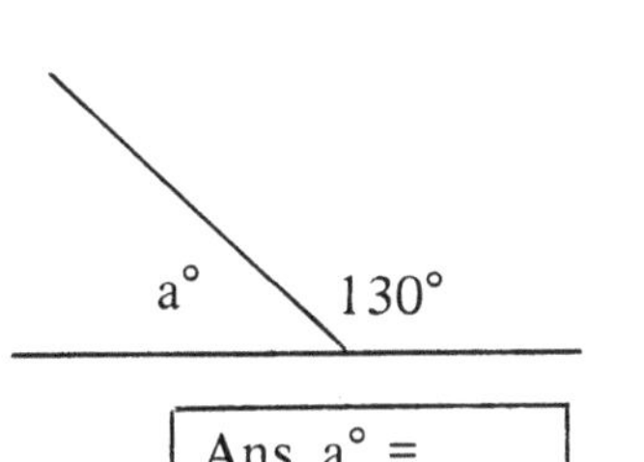

Ans. a° =

b° 37°

Ans. b° =

73° c°

Ans. c° =

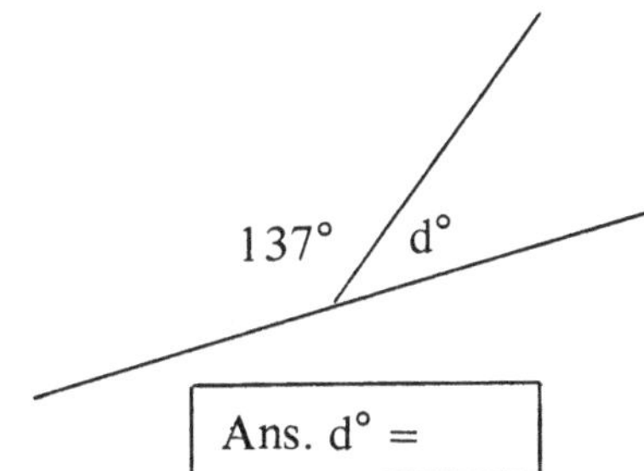

Ans. d° =

e° 84°

Ans. e° =

f° 118°

Ans. f° =

Two angles whose magnitudes sum to 180° (Two Right Angles) are called Supplementary Angles.
eg. 50° is the supplement of 130° and 130° is the supplement of 50°.

Practice Examples:

1. In each caSe, write down the supplement of the angle stated.

(a) 70° Ans. ( ) (b) 127° Ans ( ) (c) 34° Ans ( )

(d) 171°Ans. ( ) (e) 99° Ans. ( ) (f) 89° Ans ( )

2. State in the box provided, whether the following statements are true or false.

(a) 70° and 110° are supplementary

(b) 181° and 9° are supplementary

(c) 37° is the supplement of 143°

(d) 1° is the supplement of 79°

(e) 45° is not the supplement of 135°

(f) 177° is the supplement of 3°

Sum of the Magnitude of "Angles On a Line"

In each case, calculate the size of the angles marked with pronumeral. Do not use your protractor

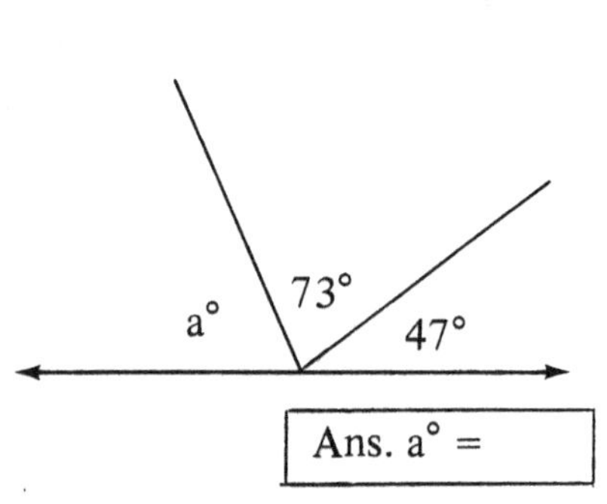

Ans. a° =

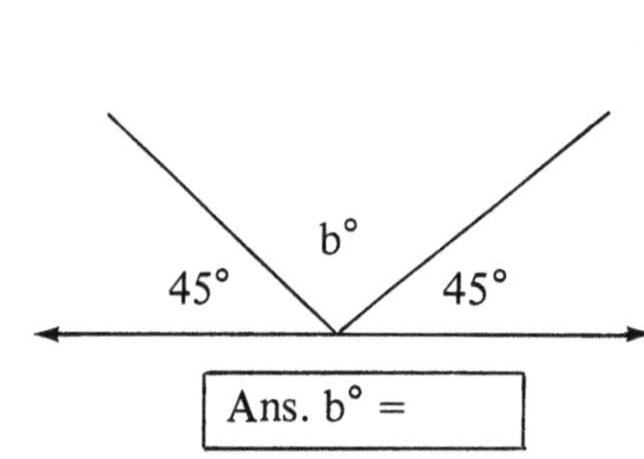

Ans. b° =

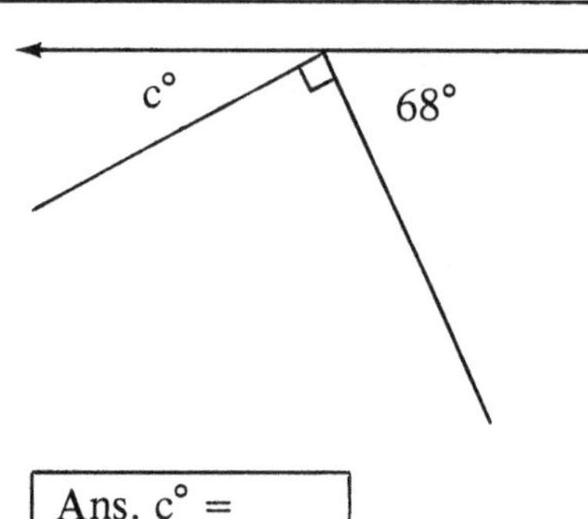

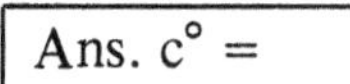

Ans. c° =

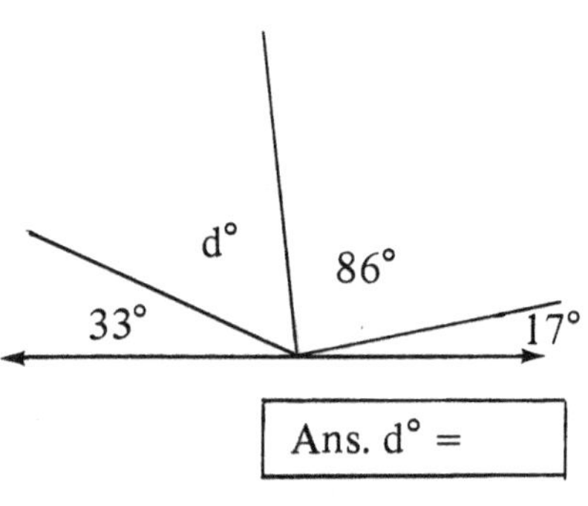

Ans. d° =

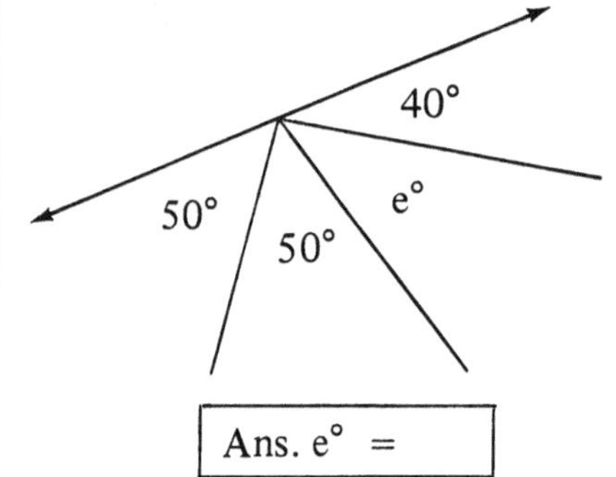

Ans. e° =

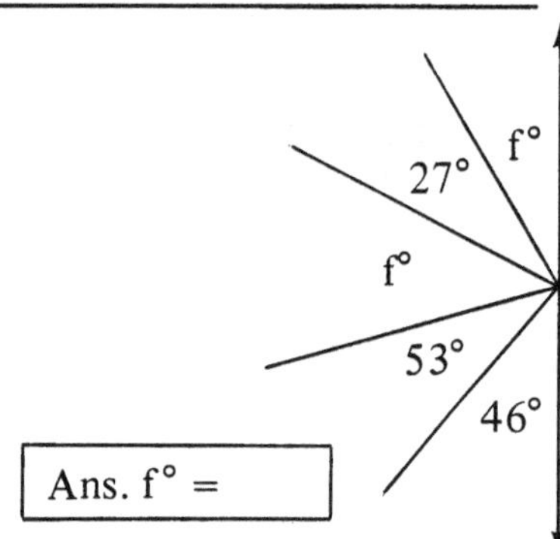

Ans. f° =

The sum of the magnitude of angles at a point on a line is equal to 180 degrees or two right angles.

In the following diagram, use your Mathomat protractor to verify the above statement, by accurately measuring the angles marked with a pronumeral. Complete the working shown.

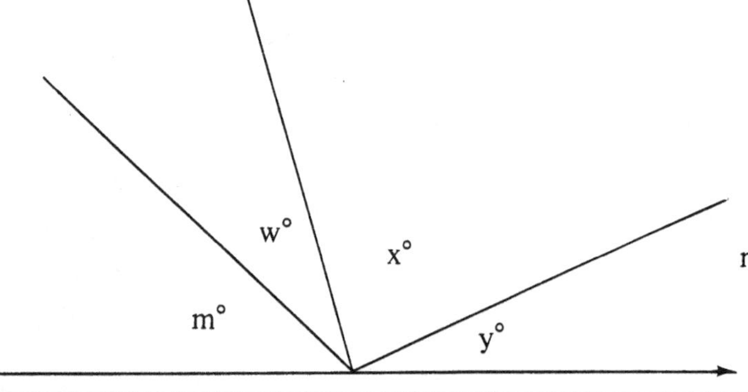

m° =

w° =

x° =

y° =

m° + w° + x° + y° =

25

Sum of the Magnitude of Angles at a Point

In each case, calculate the size of angles marked with a pronumeral. Do not use your protractor.

1. 130° 70° a°

Ans. a°= °

2. b° b° 150°

Ans.

3. 145° c°

Ans.

4. 98° 122° d° 110°

Ans.

5. 70° e° 106° 152°

Ans.

6. f° 111° 75° f°

Ans.

7. 92° 71° g° 31°

Ans.

8. h° 70° h°

Ans.

9. 30° k° 36° 87° 28° 127°

Ans.

> The sum of the magnitude of the angles at a point is equal to 360 degrees or four right angles.

In the following diagram, use your Mathomat protractor to verify the above statement, by accurately measuring the angles marked with a pronumeral. Complete the working shown.

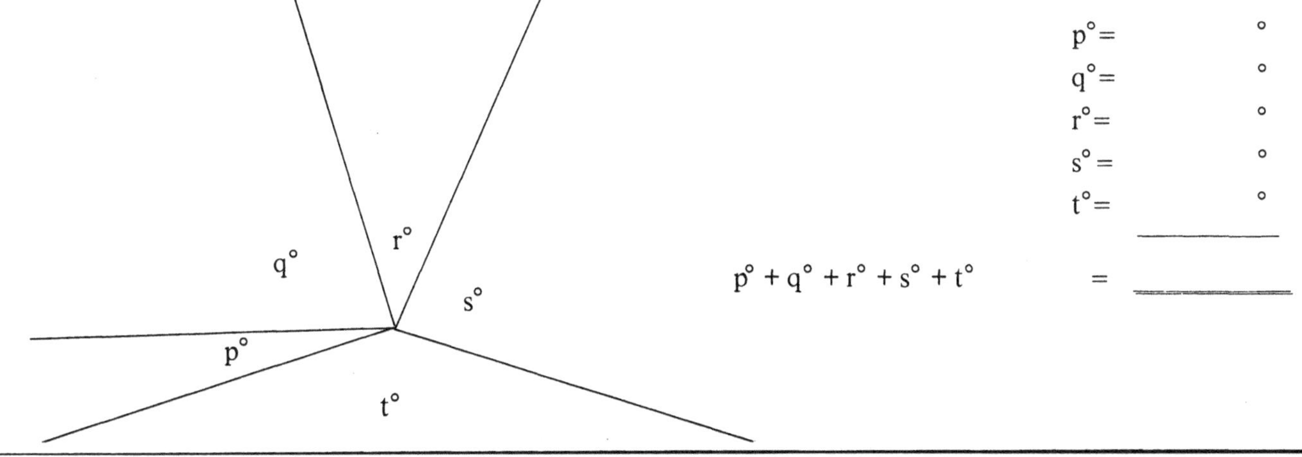

p°= °
q°= °
r°= °
s°= °
t°= °

p° + q° + r° + s° + t° = ______

VERTICALLY OPPOSITE ANGLES

In the following diagram, use your Mathomat protractor to measure the angles marked with a pronumeral.

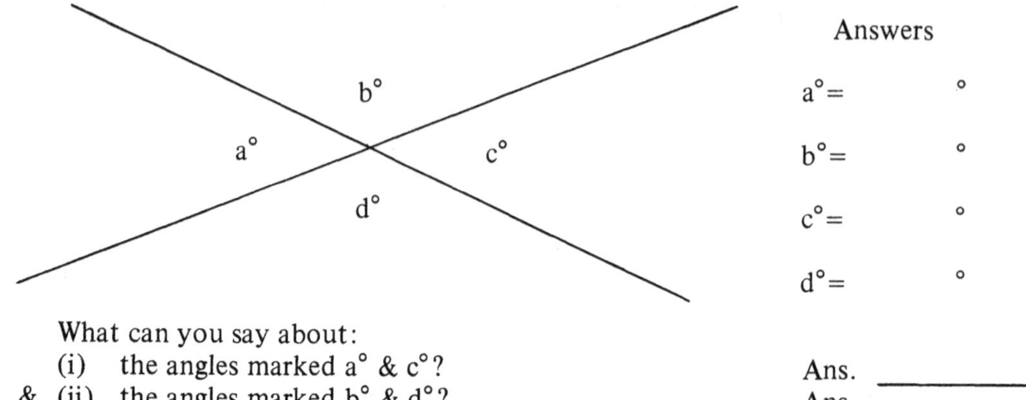

Answers

a°= °

b°= °

c°= °

d°= °

What can you say about:

(i) the angles marked a° & c°? Ans. ______

& (ii) the angles marked b° & d°? Ans. ______

The angles marked a° & c° and b° & d° are called Vertically Opposite Angles.

> If two lines intersect the Vertically Opposite Angles so formed are ______

Practice Examples:

In each case calculate the value represented by the pronumeral.

Do not use your protractor as the diagrams have not been drawn accurately.

1

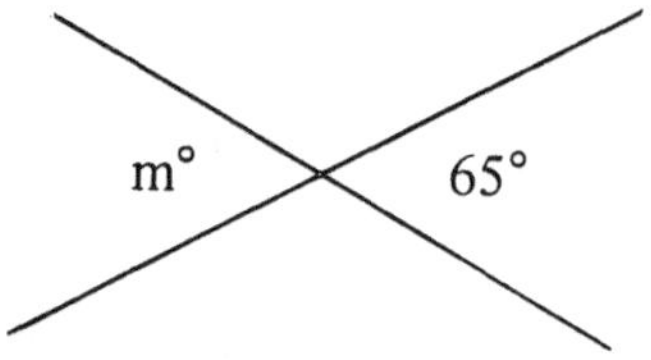

m° 65°

Ans.

2

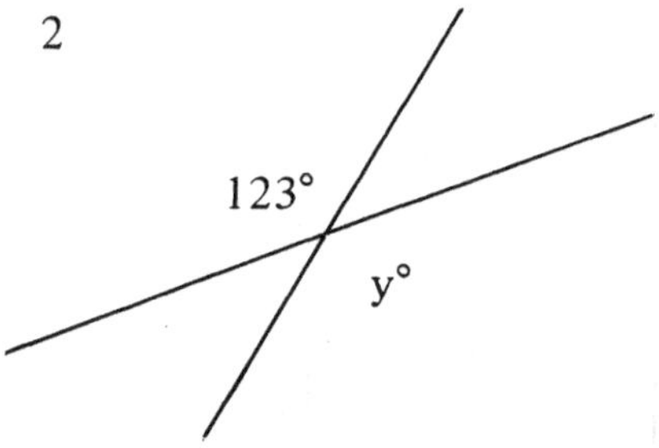

123° y°

Ans.

3

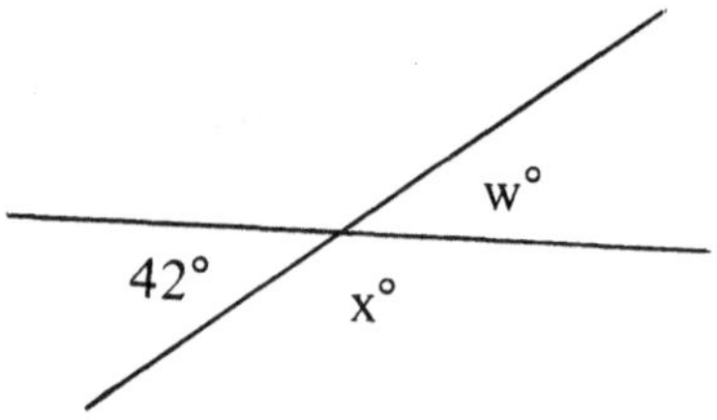

w° 42° x°

Ans.

4

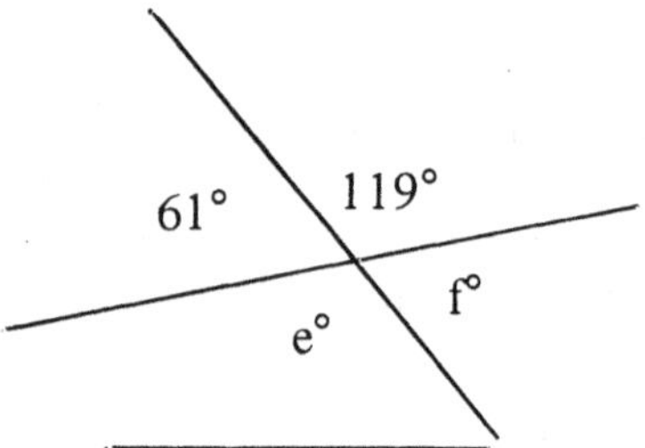

61° 119° e° f°

Ans.

5

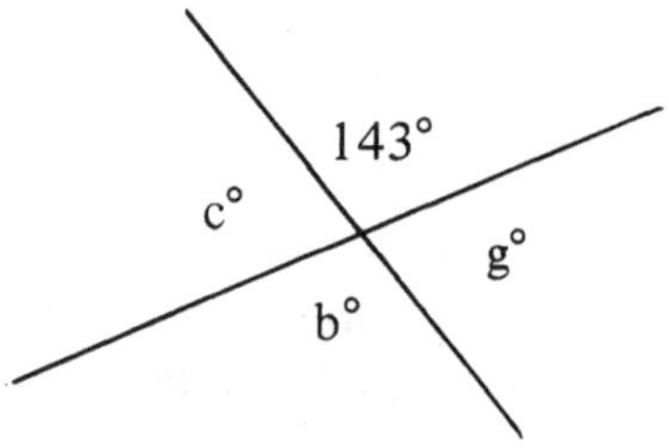

143° c° g° b°

Ans.

6

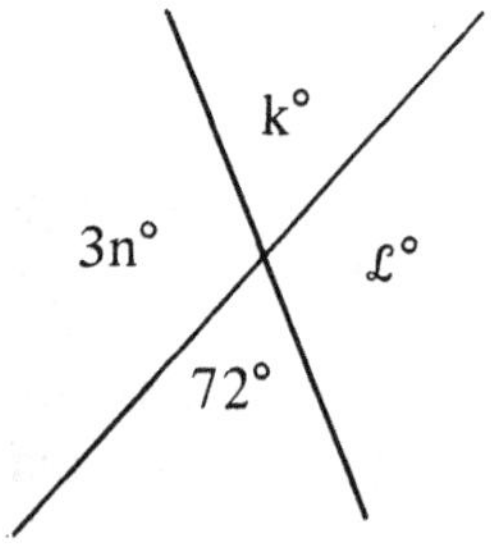

k° 3n° ℒ° 72°

Ans

7

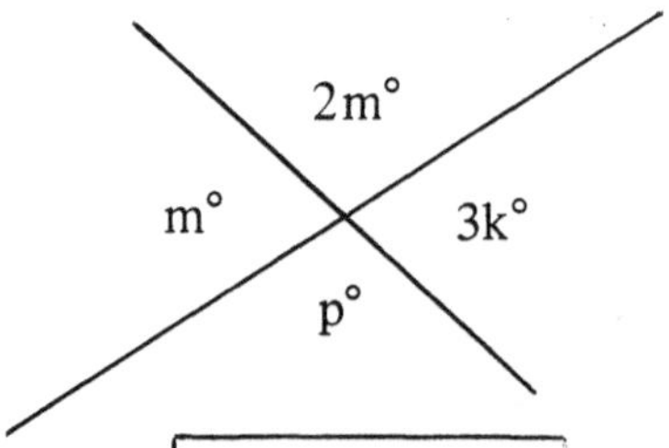

2m° m° 3k° p°

Ans.

8

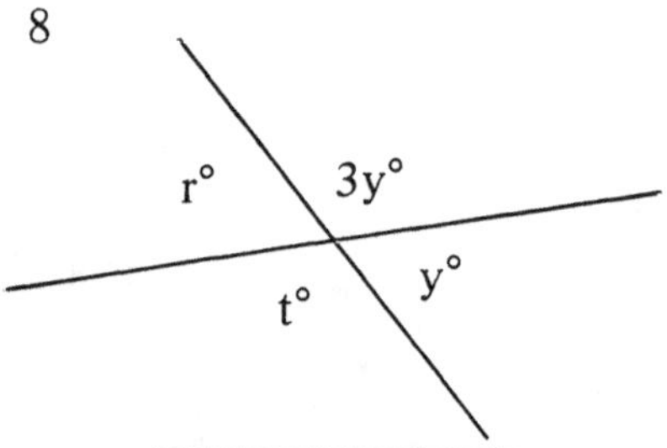

r° 3y° t° y°

Ans.

9

2b° 78° q° a°

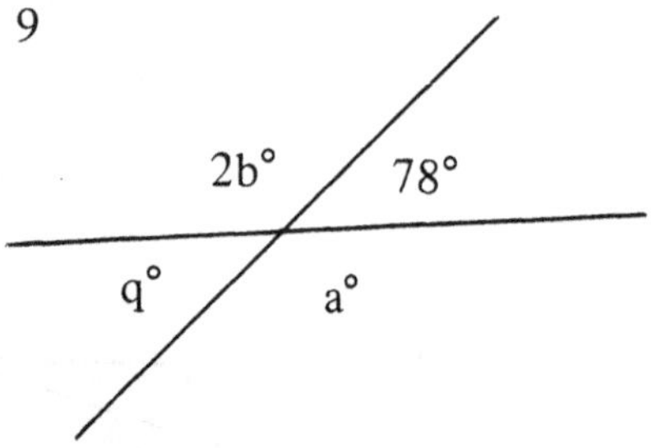

Ans.

TEST 1. **ANGLES**

Name: ____________________
Section

1. Complete this definition of an angle.
An angle is the union of two ________ with a ____________ end point.

2. What is meant by the word "magnitude" in the study of angles.
Ans. ________

3. Complete:
The magnitude of an angle is the amount of ________ that has taken place between the ________ of the angle.

4. Write down 3 things that turn, revolve, rotate or spin.
(i) ____________________
(ii) ____________________
(iii) ____________________

5. What is the mathematical name for one complete turn?
Ans. ____________

6. The following lines have been drawn at right angles. Suitably mark each right angle

(a) (b)

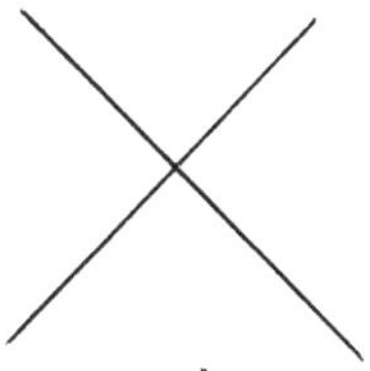

(c) 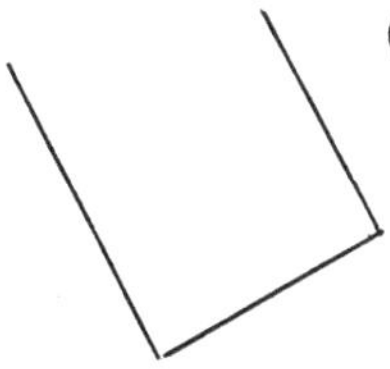 (d)

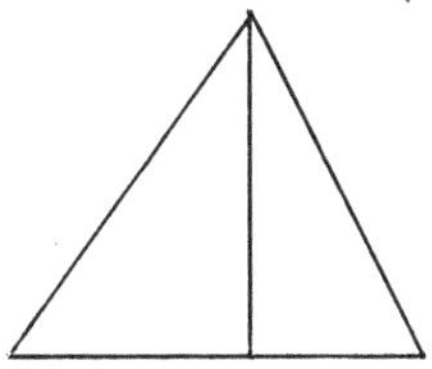

7. Complete this definition of perpendicular lines.
Perpendicular lines are lines that meet or ____________ each other at ____________

8. Complete:
The common end point of an angle is called the ____________ of the angle.

9. How many right angles are there in
(a) ¾ of one complete turn? [ ]
(b) 1½ complete turns? [ ]
(c) 3¼ complete turns? [ ]

10. In each case name (in the space provided) the type of angle shown below.
(a) Ans. [ ] (b) Ans. [ ]
(c) Ans. [ ] (d) Ans. [ ]
(e) Ans. [ ] (f) Ans. [ ]

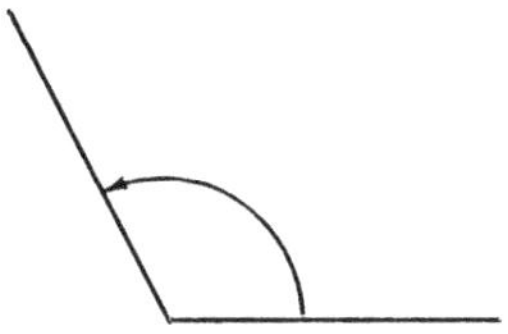

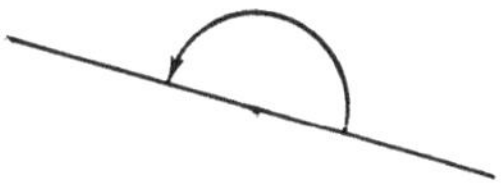

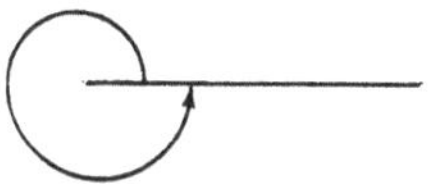

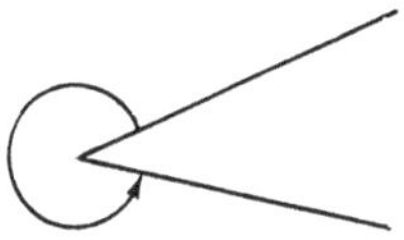

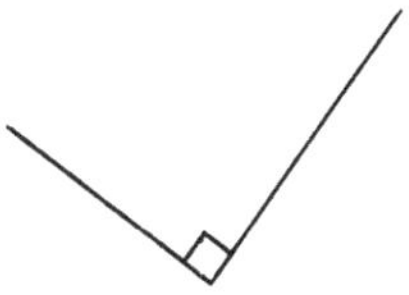

11. How many degrees are there in:
(a) 1 revolution? Ans. ______
(b) $^1/_6$ of a revolution? Ans. ______
(c) $^1/_8$ of a revolution? Ans. ______
(d) $^1/_{12}$ of a revolution? Ans. ______

12. In each case write down the complements of
(a) 37° Ans. ______
(b) 89° Ans. ______
(c) 61° Ans. ______
(d) 8° Ans. ______

13. How many degrees are there in
(a) 2 right angles? Ans. ______
(b) 7 right angles? Ans. ______

14. In each case write down the supplement of
(a) 150° Ans. ______
(b) 68° Ans. ______
(c) 100° Ans. ______
(d) 93° Ans. ______

TEST 2 **ANGLES** Name __________________
Section

1. Using your Mathomat protractor to accurately mesure the following angles.

(a)

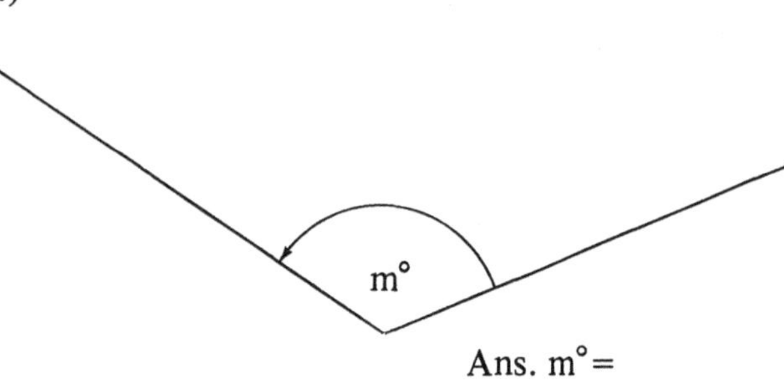

Ans. m°=

(b)

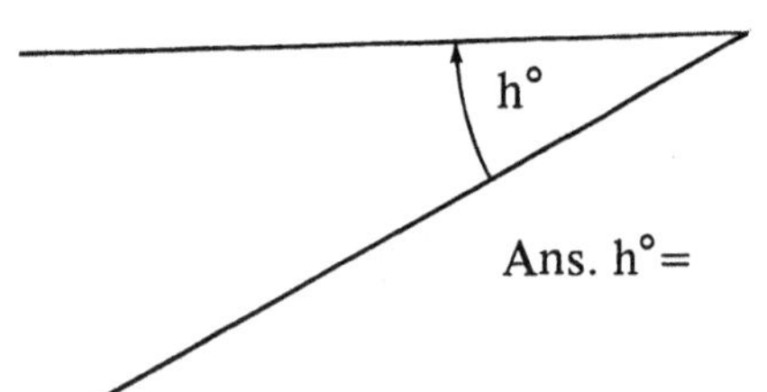

Ans. h°=

(c)

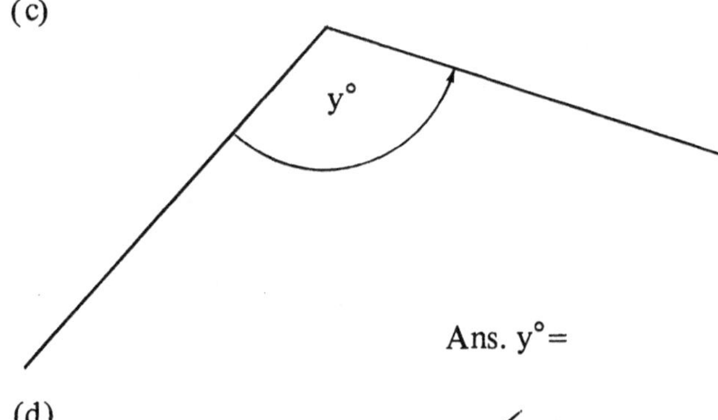

Ans. y°=

(d)

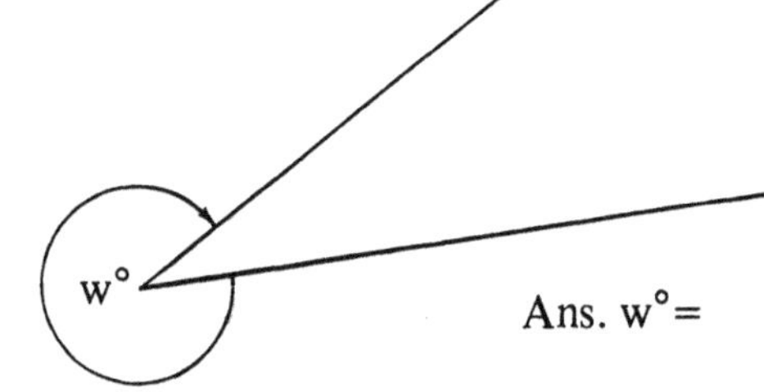

Ans. w°=

(e)

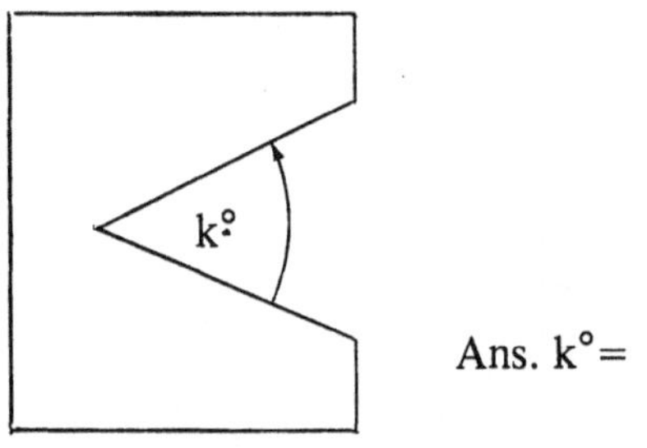

Ans. k°=

2. Using your Mathomat protractor and the space provided, draw the following angles. Clearly indicate the angles.

(a) 147°

(b) 315°

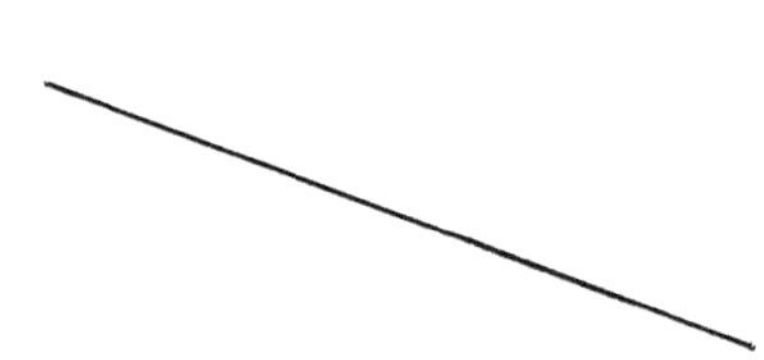

3. In each case calculate the value represented by the pronumerals (letters).

(a)

Ans. b°=

(b)

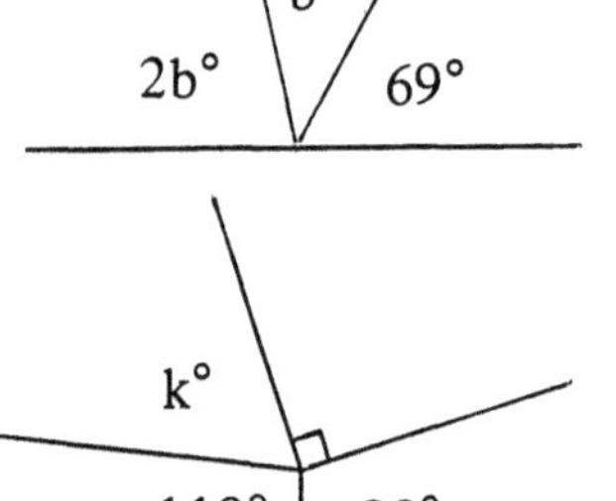

Ans. k°=

(c)

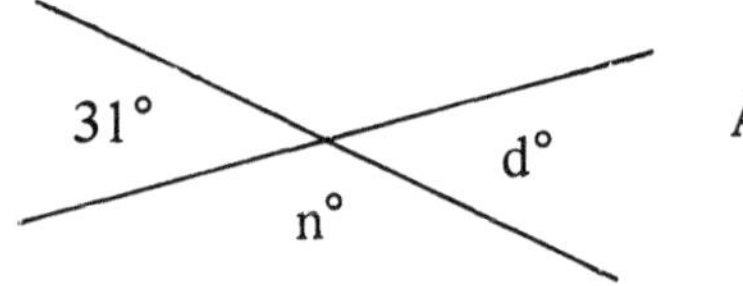

Ans. d°=
n°=

(d)

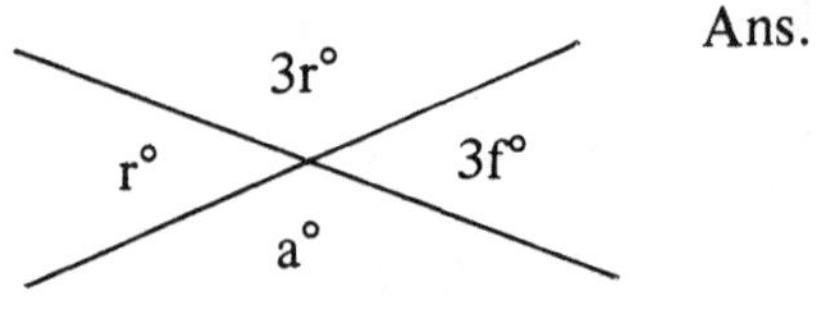

Ans. r°=
f°=
a°=

BEARINGS

Navigators and Captains of ships and aeroplanes must be able to fix and plot their positions with great accuracy.

If you are at any time thinking of participating in or joining any one of the following activities or occupations, you too will need to know how to fix positions and accurately describe the direction of your movements.

Car or bike rallies
Boating or yatching
Hiking or bush walking
Learning to fly a light aircraft or glider
The Police or Civil Ambulance – Search & Rescue Squad
Scouts or Girl Guides
Surveying
Fire Brigade
The Army, Navy or Air Force

The following are two methods that enable us to describe and measure aq particular direction. Both involve using a magnetic compass.

They are:

(A) Cardinal Point Method.
(B) 360° Bearing Method.

Any direction given by one of the above methods is called a <u>Bearing.</u>

(A) <u>Cardinal Point Method</u>

Consider the drawing of the compass shown or refer to Fig.23 of your Mathomat.

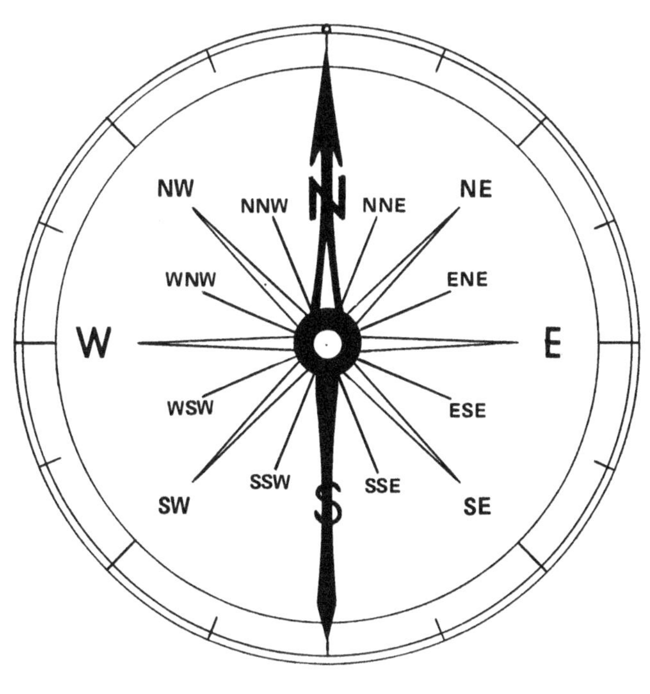

Here the 16 Cardinal Points are marked around the circumference of a circular card.

The North Point (N) of of the compass card is established by the magnetized needle pointing to the magnetic North Pole.

The compass card is divided into quarters from this point, obtaining the points East (E), South (S), West (W).

The card is further divided into eighths obtaining the points NE, SE, SW, NW.

Then again into sixteenths obtaining the points NNE, ENE, ESE, SSE, SSW, WSW, WNW, NNW,

Below are some symbols used to indicate a North direction.

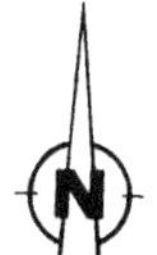

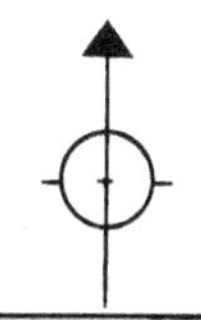

BEARINGS

The Cardinal Point Method is used mainly when only general directions or bearings are involved, that is when accuracy is not essential.

Examples:

(a) The sun rises due East ands sets due West.
(b) That plane is flying due South.
(c) The hurricane came in across the coast froma NNE direction.
(d) the water tower from here is in a South Easterly direction.
(e) Strong Northerly winds are expected tomorrow,
(f) The spin of the Earth is from West to East.
(g) The patrol boat was travelling in a North Westerly direction.

Practice Examples:

1. Use Fig. 23 of your Mathomat to help answer the following.

   In each case, write dwon the compass direction of positions A to H from point P.

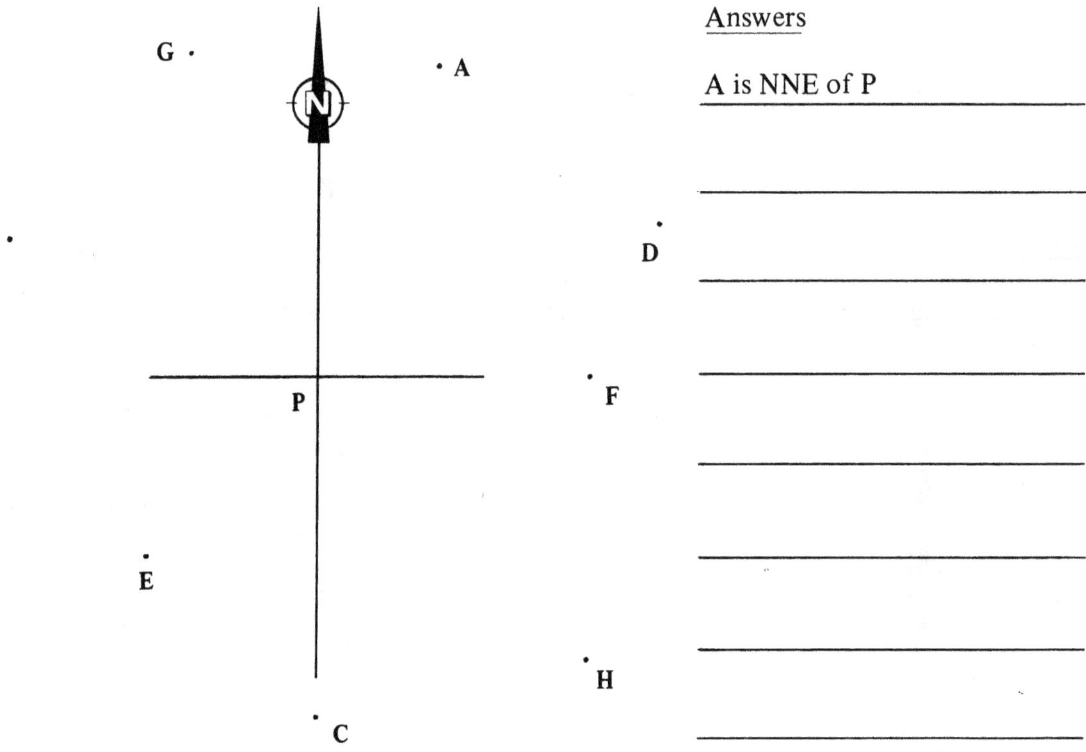

Answers

A is NNE of P ______

______

______

______

______

______

______

______

.2 Use your Mathomat protractor to find the angle in degrees between the North direction and the direction of:

(i) E or W Ans. ______
(ii) NE or NW Ans. ______
(iii) NNW or NNW Ans. ______
(iv) ENE or WNW Ans. ______

The South direction and the direction of:

(i) E or W Ans. ______
(ii) NE or NW Ans. ______
(iii) NNW or NNW Ans. ______
(iv) ENE or WNW Ans. ______

TREASURE HUNT

3. The treasure of gold coins and precious gems is very cunningly hidden under one of a large number of rocks marked A,B,C,D,E,F,G on the map. These rocks are well concealed and can only be found by following these directions. From "Black-Jack" lagoon, proceed 7,000 paces in a SSE direction to the bank of the crooked river, then due East along the left bank for 8,500 paces, then due North for 4,500 paces, then WNW for 5,500 paces, then NE until you strike the coast to the North of a group of pine trees, then due West for 3,000 paces, then SW for 2,500 paces. A cluster of large rocks should then be visible. The rock directly in line with the last bearing is the one under which Captain Blood's treasure is buried.

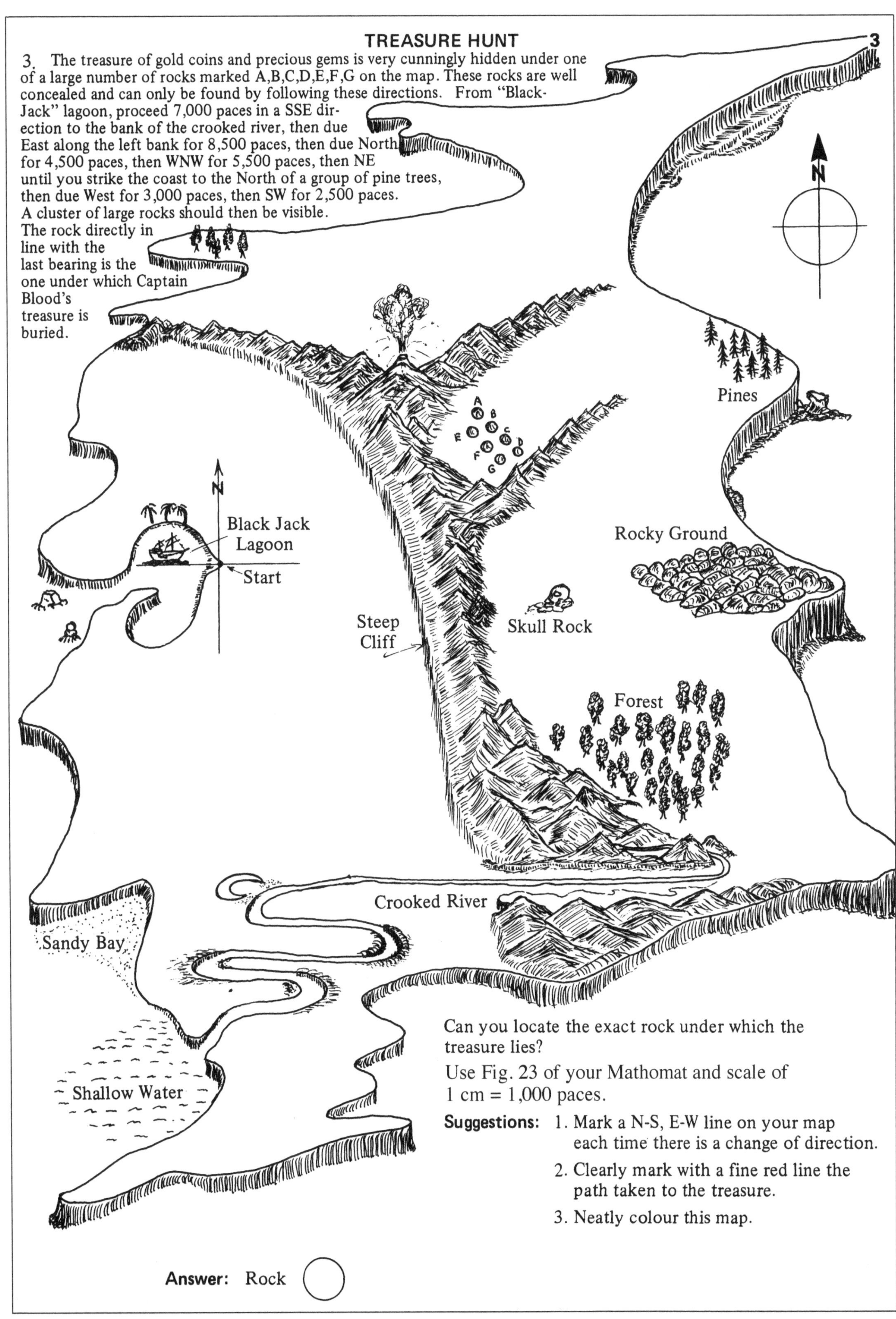

Can you locate the exact rock under which the treasure lies?

Use Fig. 23 of your Mathomat and scale of 1 cm = 1,000 paces.

Suggestions:
1. Mark a N-S, E-W line on your map each time there is a change of direction.
2. Clearly mark with a fine red line the path taken to the treasure.
3. Neatly colour this map.

Answer: Rock ◯

BEARINGS

(B) 360° Bearing Method

As shown below, a modern compass has not only 16 cardinal points but the 360° of a circle marked on it, starting with 0 at north, 90 at east, 180 at south and 270 at west.

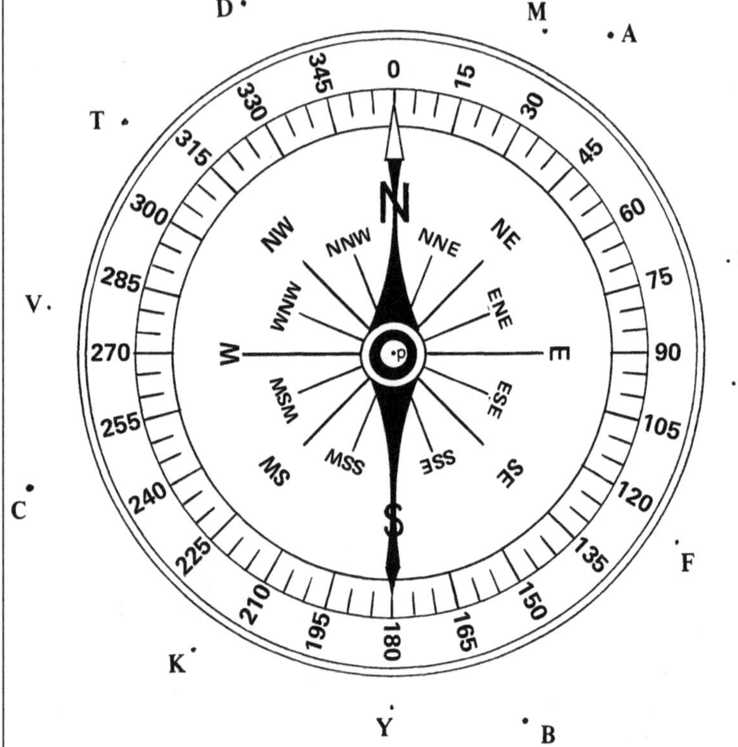

Using this method the bearing of an object from P is always measured in a <u>clockwise</u> direcion from <u>North.</u>

Examples:

(i) the bearing of A from P is 035°

(ii) the bearing of B from P is 159°

(iii) the bearing of C from P is 250°

(iv) the bearing of D from P is 337°

Because of its simplicity and accuracy, navigators throughout the world use this method to give a bearing or course.

Note:

In statingthe bearingof A from P a navigator would say zero-three-five degrees (Magnetic).

Exercises:

1. Write in figures and state in words how a navigator would say the bearings of the following points from P (above). Suggestion: Draw a light line from the centre of the cpmpass to each lettered point.

(a) M 027° M zero-two seven degrees (Magnetic)

(b) T ______________________________

(c) K ______________________________

(d) R ______________________________

(e) Y ______________________________

(f) H ______________________________

(g) V ______________________________

(h) F ______________________________

BEARINGS

2 Study the map of the locality shown.

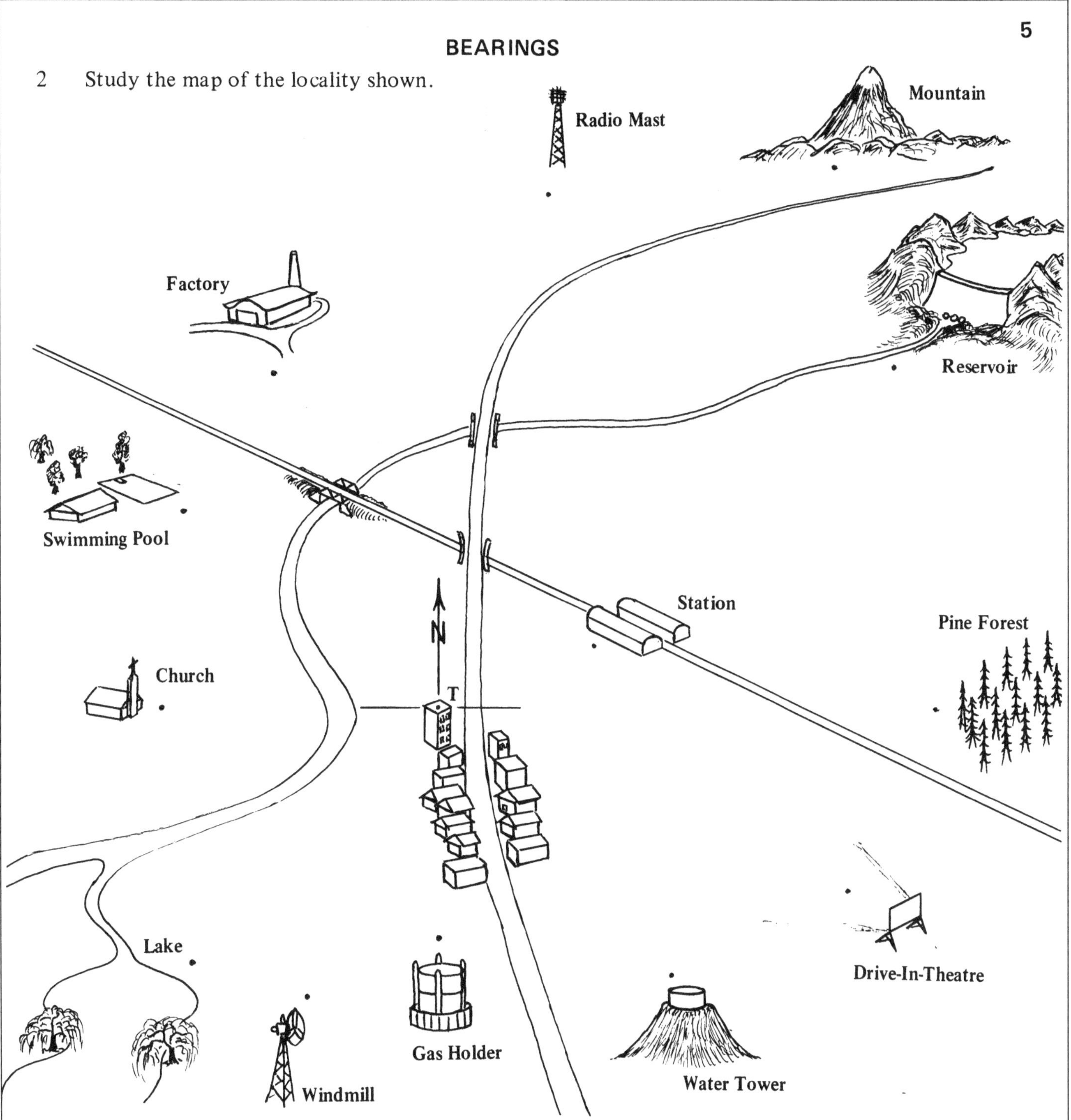

Now imagine your are stnading on the roof of the tall building (T). In each case, draw a straight line from T tothe dot near each object, and using your Mathomat protector, give its bearing.

Answers

1.	Radio Mast	012° Mag.	7.	Factory	______
2.	Windmill	______	8.	Gas Tower	______
3.	Pine Forest	______	9.	Reservoir	______
4.	Drive-In Theatre	______	10.	Lake	______
5.	Church	______	11.	Water Tower	______
6.	Station	______	12.	Mountain	______
			13.	Swimming Pool	______

BEARINGS

3. In each case, using point 0 below and your Mathomat, draw lines to indicate the following bearings. Clearly mark the angle turned through.

(a) A Mirage Jet flying on a course 073°

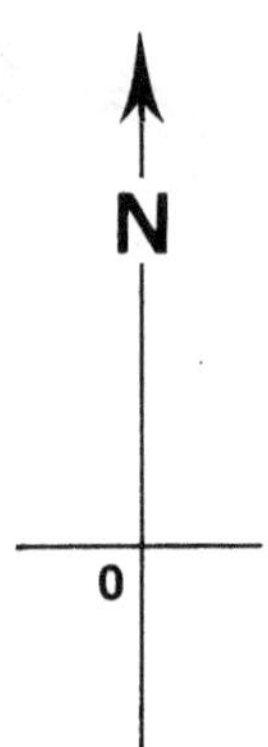

(b) A passenger liner moving on a course 158°

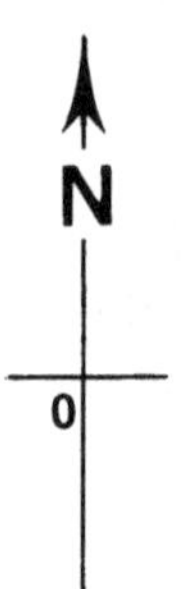

(c) A helicopter flying on a course 270°

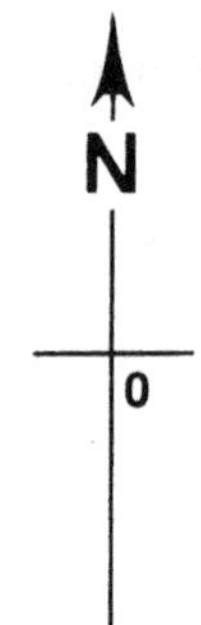

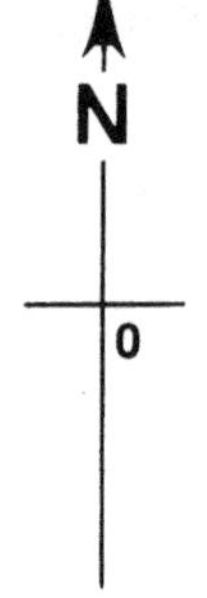

(d) A submarine cruising on a course 252°

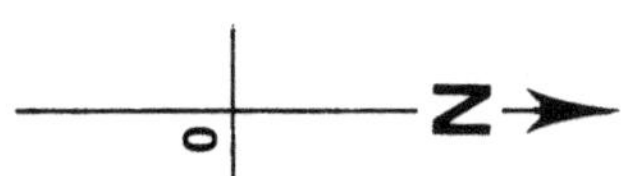

(e) A rocket fired on a course 338°

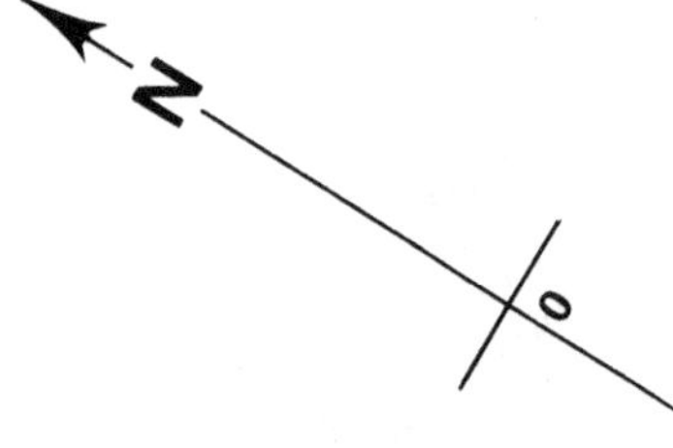

(f) A Boeing 707 Jet on a course 200°

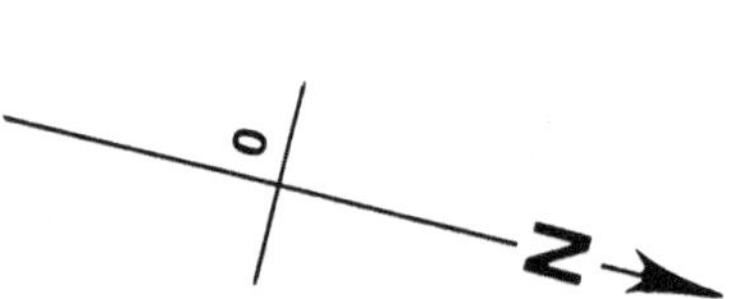

4 Write down the final bearing of each of the following:

(a) Flying on a bearing of 047°, then made a 67° left turn.
(Ans. ________)

(b) Sailing on a bearing at 310°, then turned 85° to port.
(Ans. ________)

(c) Flying on a bearing of 103°, then made a 210° right turn.
(Ans. ________)

(d) Sailing on a bearing of 245°, then turned 156° to starboard.
(Ans. ________)

(e) Flying on a bearing of 195°, made a 265° left turn.
(Ans. ________)

2. In each case, using a scale of 1 cm = 100 km and your Mathomat protractor, describe the following courses from A to D in a similar manner to the worked example below.

Example:

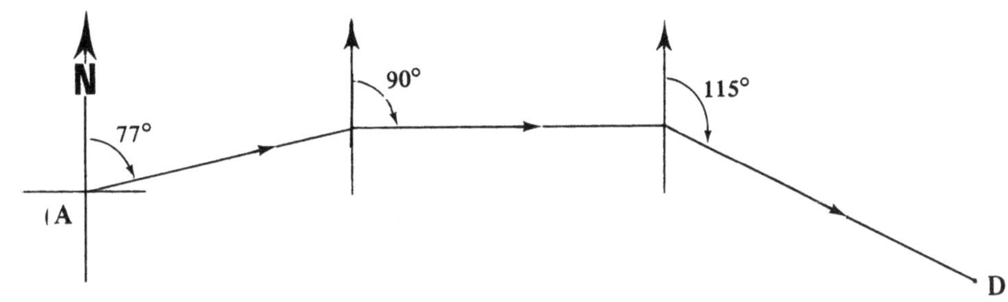

Answer: From A 077° (mag.) for 350km then 090° for 400km then 115° for 450km.

(a)

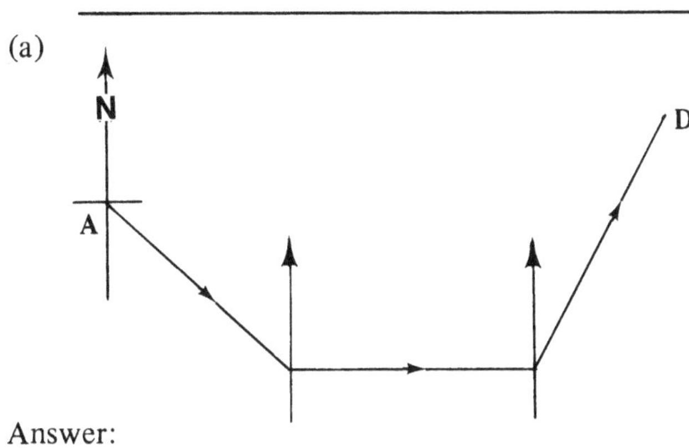

Answer:

(b)

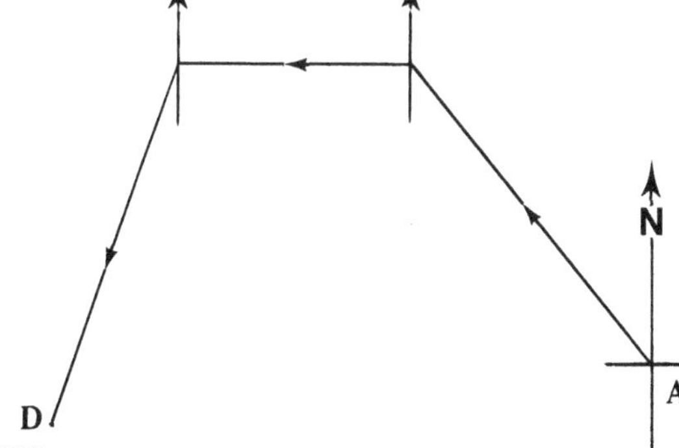

Answer:

(c)

Answer:

(d)

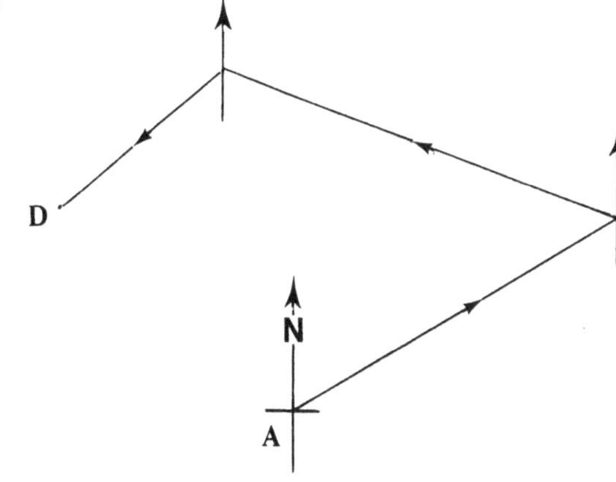

Answer:

(e)

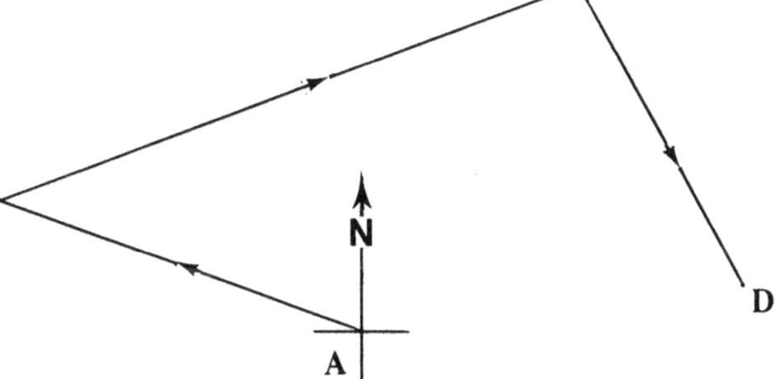

Answer:

(f)

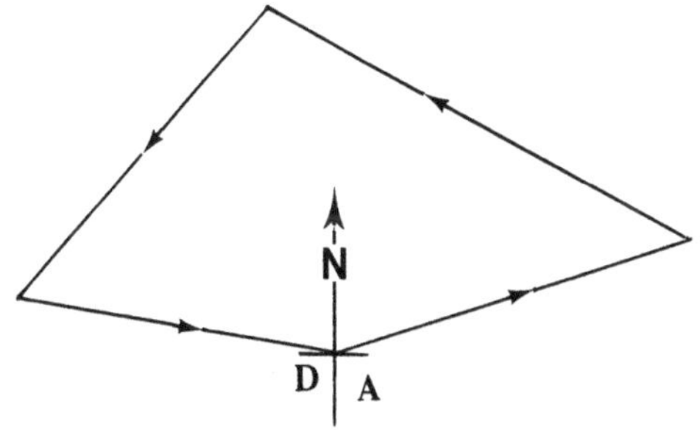

Answer:

BEARINGS

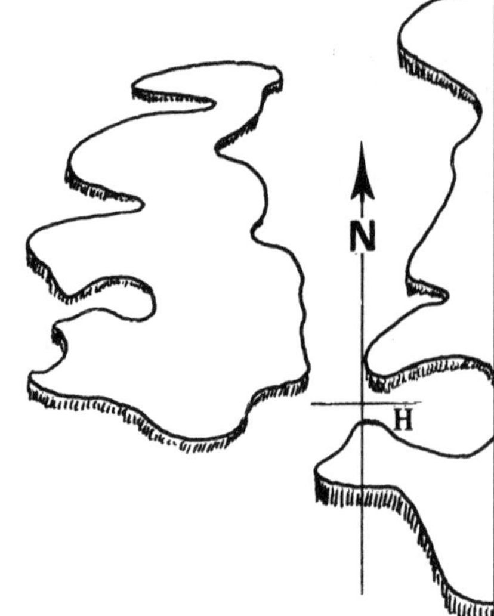

3. A nuclear powered submarine sets out from its home base H on a reconnaissance exercise and moves completely submergede along the following course; 348 (mag) for 500 km, then 237 for 550km, then 140° for 500km.

Using a scale of 1cm = 100km and your Mathomat protractor, plot this course in the space provided opposite and write down the bearing and distance of the home base H from the submarine.

Ans.

4. In each case, calculate the bearing of H from P. Do not use your protractor as the diagrams have not been drawn accurately.

(a)

Ans.

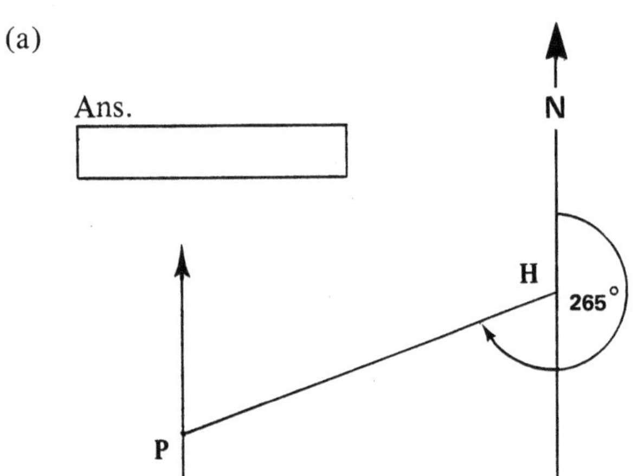

(b)

Ans.

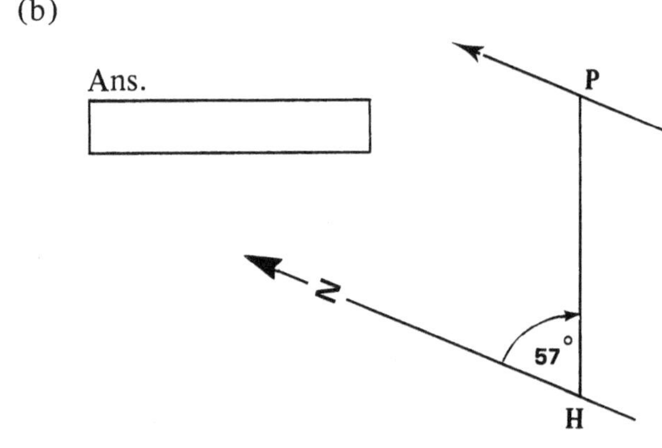

5. A boeing 707 Passenger Jet and a Jumbo Jet leave an airport at the same time. The 707 flies on a bearing of 296° at 1000km per hr. While the Jumbo flies on a bearing of 045° at 700km per hr.

Using your Mathomat protractor, a scale of 1cm - 100km and the same diagram, show their positions and distances apart after (a) 1/2 hour (b) 1 hour.

Write down in the space below, their distance apart after 1 hour.

N

Airport

Answer:

Distance apart after 1 hour is ________ km

Plotting a Course

Example: A pilot of a light aircraft described his journey from the aerodrome to a private landing field in the following way. From the drome at a height of 700 metres, I flew 073° (magnetic) for 250km to Lake Taupo, then 128° for 300km to Knobby's Peak, then 098° for 200km to Joe's field. When landing, a number of quail hit the prop and window, but fortunately no damage was done.

Using a scale of 1cm = 50km and your Mathomat protractor, plot his course.

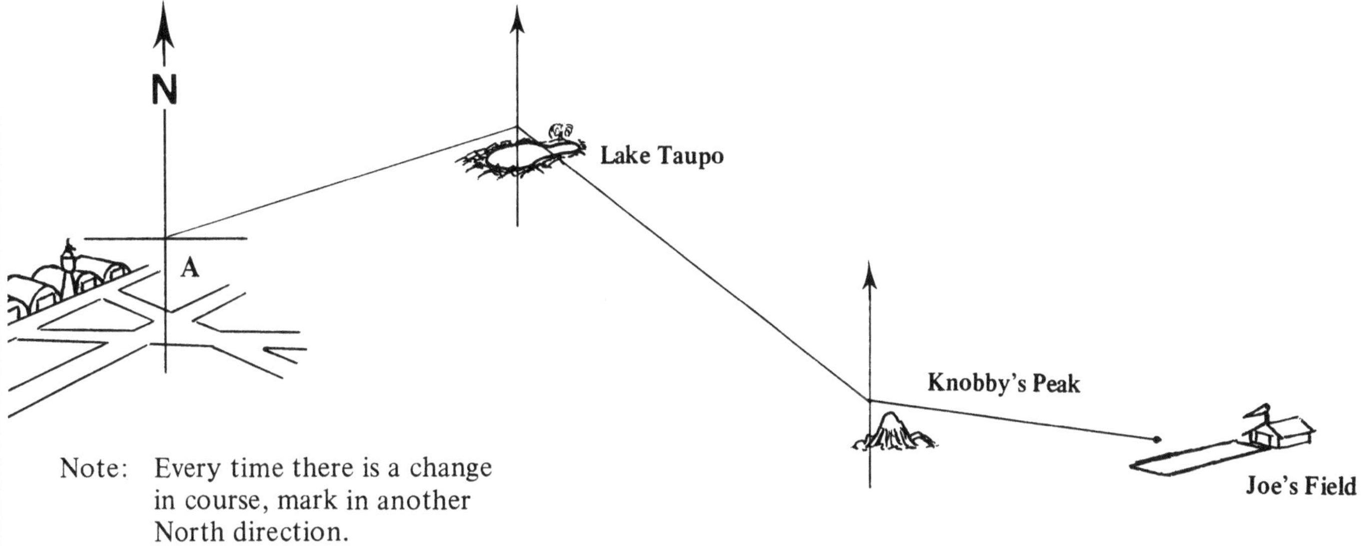

Note: Every time there is a change in course, mark in another North direction.

Practice Examples:

1 In each case, using a scale of 1cm = 100km and your Mathomat protractor, plot the following courses from the aerodrome A to your destination D.

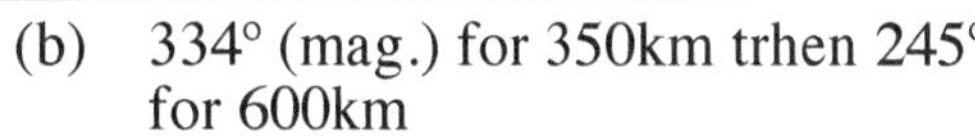

(a) 082° (mag) for 500km then 130° for 300km

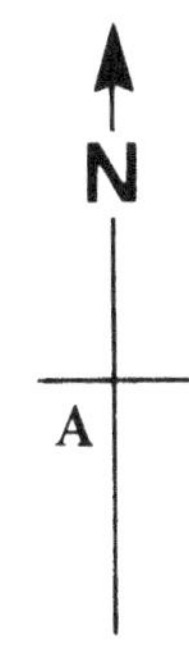

(b) 334° (mag.) for 350km trhen 245° for 600km

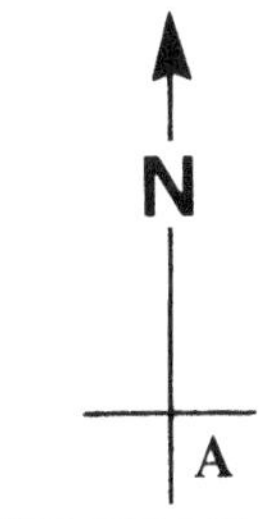

(c) 067° (mag.) for 700km then 142° for 200km then 255° for 450km

(d) 265° (mag.) for 400km then 326° for 350km then 108° for 550km.

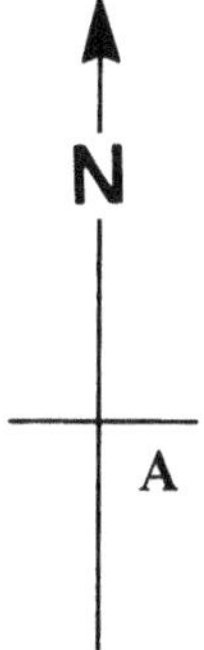

BEARINGS

if flying a light aircraft at a height of 610 metres (2000 ft.) and visibility is perfect, the distance to the horizon on 88km (55 miles). However, this visibility may be considerably reduced by either cloud, fog, dust and haze, so it is very important to check that your bearings and course are accurate and correct, particularly if flying from the mainland to a small island, as an error of 1° in a short distance of 60km (37 1/2 miles) miles could put you 1km (5/8 mile) to either side of your destination.

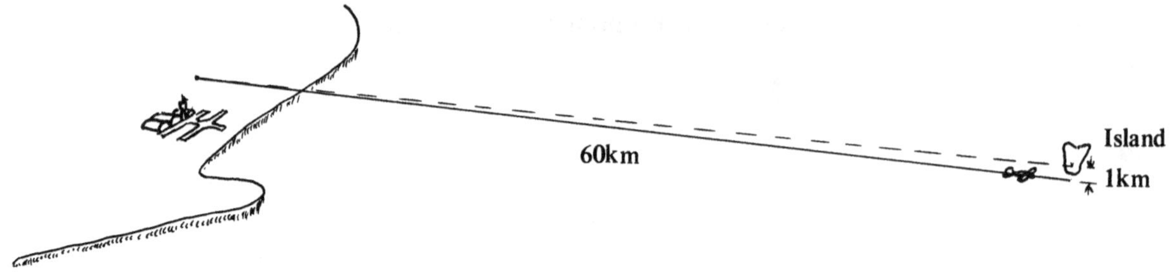

The point, in plotting his course has to make a number of adjustments in order to arrive at the correct compass bearing, They are:

1. a correction for magnetic variation that varies from place to place on the Earth's surface.
2. a compass card correction.
3. and a wind correction.

These adjustments, if not correctly applied, could put a pilot off course by as much as 25° or more, and this could be disastrous as, as mentioned above, on a short trip pf pn;y 60km (37 1/2 miles) this error would amount to 25km (20 1/2 miles) or more, and would put the pilot over unfamiliar surroundings and perhaps byond the limit of

(i) his ability to see his destination
& (ii) the plane to return to its baase.

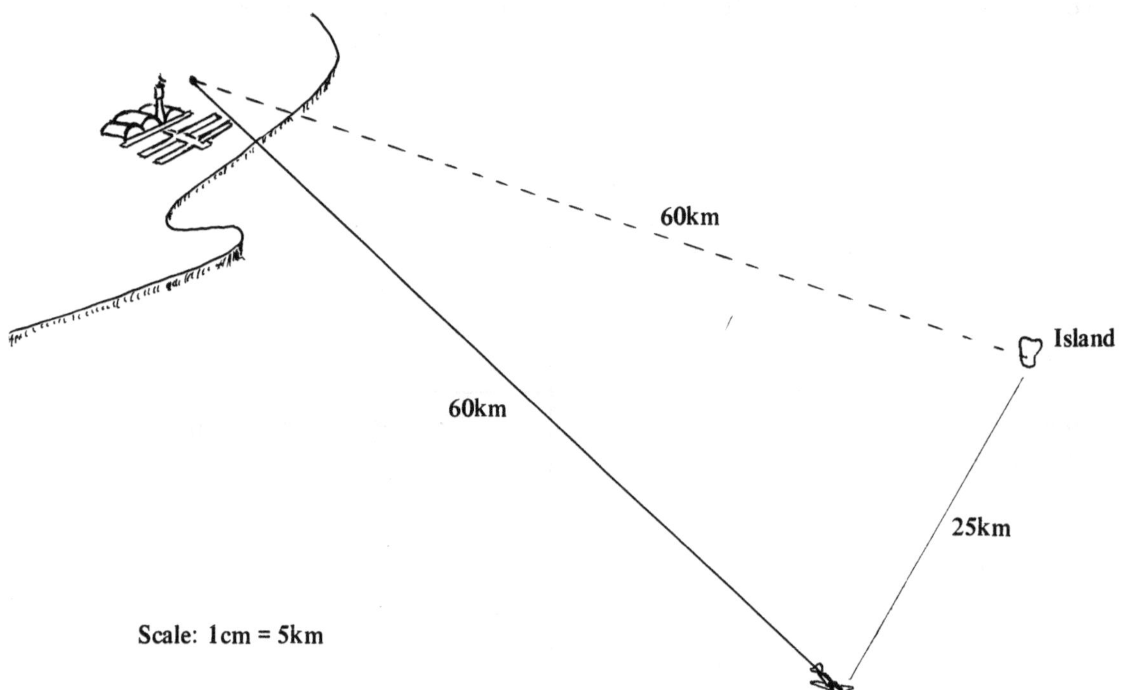

Aviation personnel, in searching for light aircraft reported missing, try to obtaqin the last reported position, bearing, and intended destination. They then send szearch planes along the misclculated course and hopefully the plane is found along this course.

RADAR

If you have visited airports or shipping ports, yo may have seen a totaing aerial on the top of the airport control tower or high up near the funnel or a large passenger liner or destroyer and wondered what they are. They are Radar aerials, and they serve a very important purpose.

In 1930-34 Radar was just being developed by scientists in America, Britain, France and Germany. This development was hastened by the growing threat of the Scond World War (1939), so much so that the British, suspecting that they would be directly involved, quickly recruited many of their best scientists and had them work togetherto invent, construct and install a chain of lage radar-warning stations on their shores.

This foresight was included to their advantage, as it gave adequate warning of the approach of enemy aircraft, enabling the small force of British fighter planes and anti-aircraft artillery to fight off the massive German air attacks by day and particularly at night.

Radar was also accredited, 1941, of detecting the Japanese air fleet approaching Pearl Harbour but the blips caused by this fleet were thought to be American bombers.

How Does Radar Work?

Most radars send out short, intense pulses of radio waves from the antenna with transmitters, and then "listen" between pulses with radio receivers for reflected waves from the object. These echoes appear as bright spots or "blips" on the radar screen, which is like the front of a television set (see diagram next page).

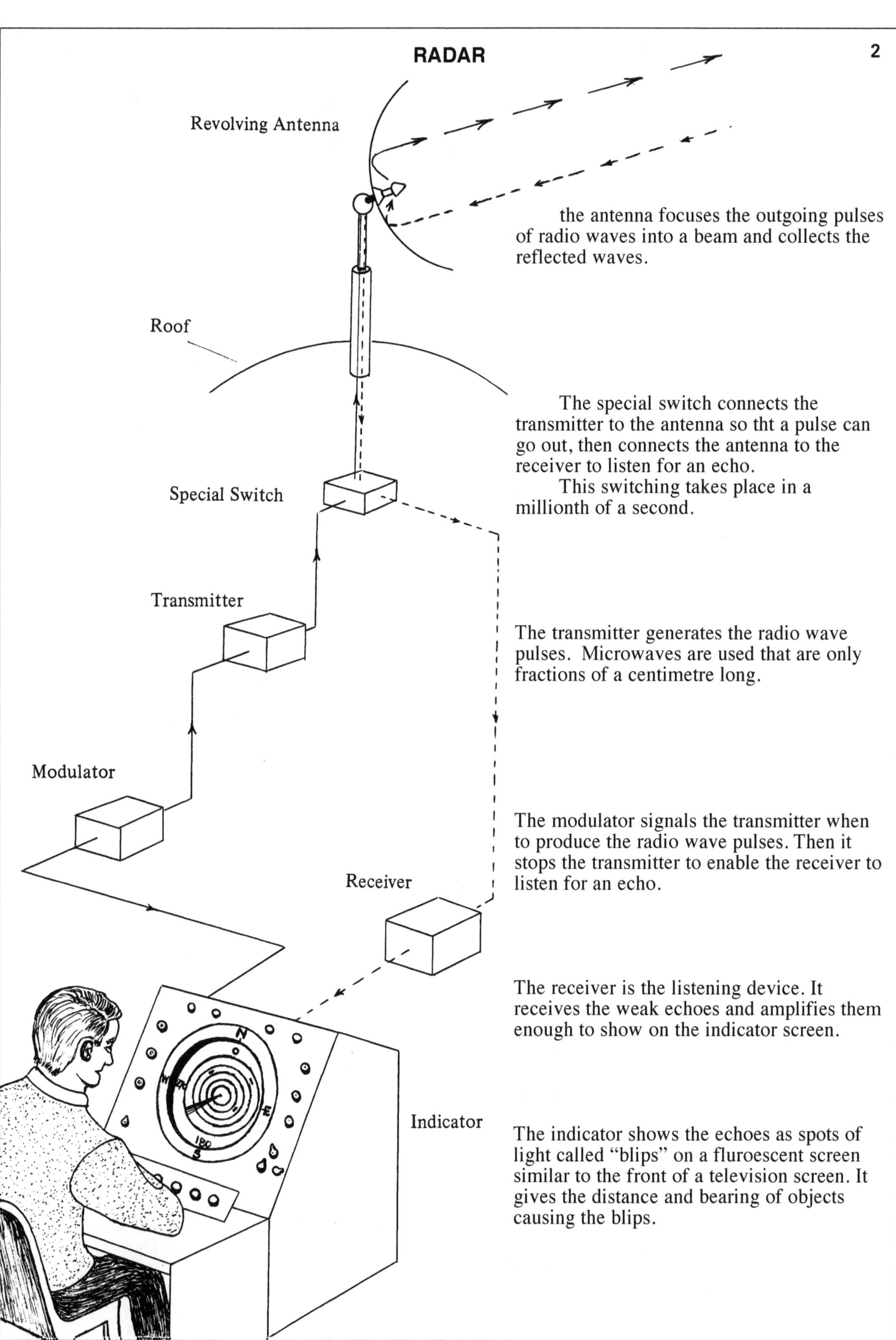
RADAR
2
Revolving Antenna
the antenna focuses the outgoing pulses of radio waves into a beam and collects the reflected waves.
Roof
The special switch connects the transmitter to the antenna so tht a pulse can go out, then connects the antenna to the receiver to listen for an echo.
This switching takes place in a millionth of a second.
Special Switch
Transmitter
The transmitter generates the radio wave pulses. Microwaves are used that are only fractions of a centimetre long.
Modulator
The modulator signals the transmitter when to produce the radio wave pulses. Then it stops the transmitter to enable the receiver to listen for an echo.
Receiver
The receiver is the listening device. It receives the weak echoes and amplifies them enough to show on the indicator screen.
Indicator
The indicator shows the echoes as spots of light called "blips" on a fluroescent screen similar to the front of a television screen. It gives the distance and bearing of objects causing the blips.

RADAR

As th antenna rotates, a thin line of light clled the "trace" sweeps around the screen in a clockwise direction. When th trace passes through a blip, both the direction and distance of the object causing the blip cn be found on the screen.

This is achieved by making the screen with concentric circles. These give distances in kilometres from the centre. A compass scale runs around the edge of the screen, and this gives the bearing of the ogject. (See diagram below).

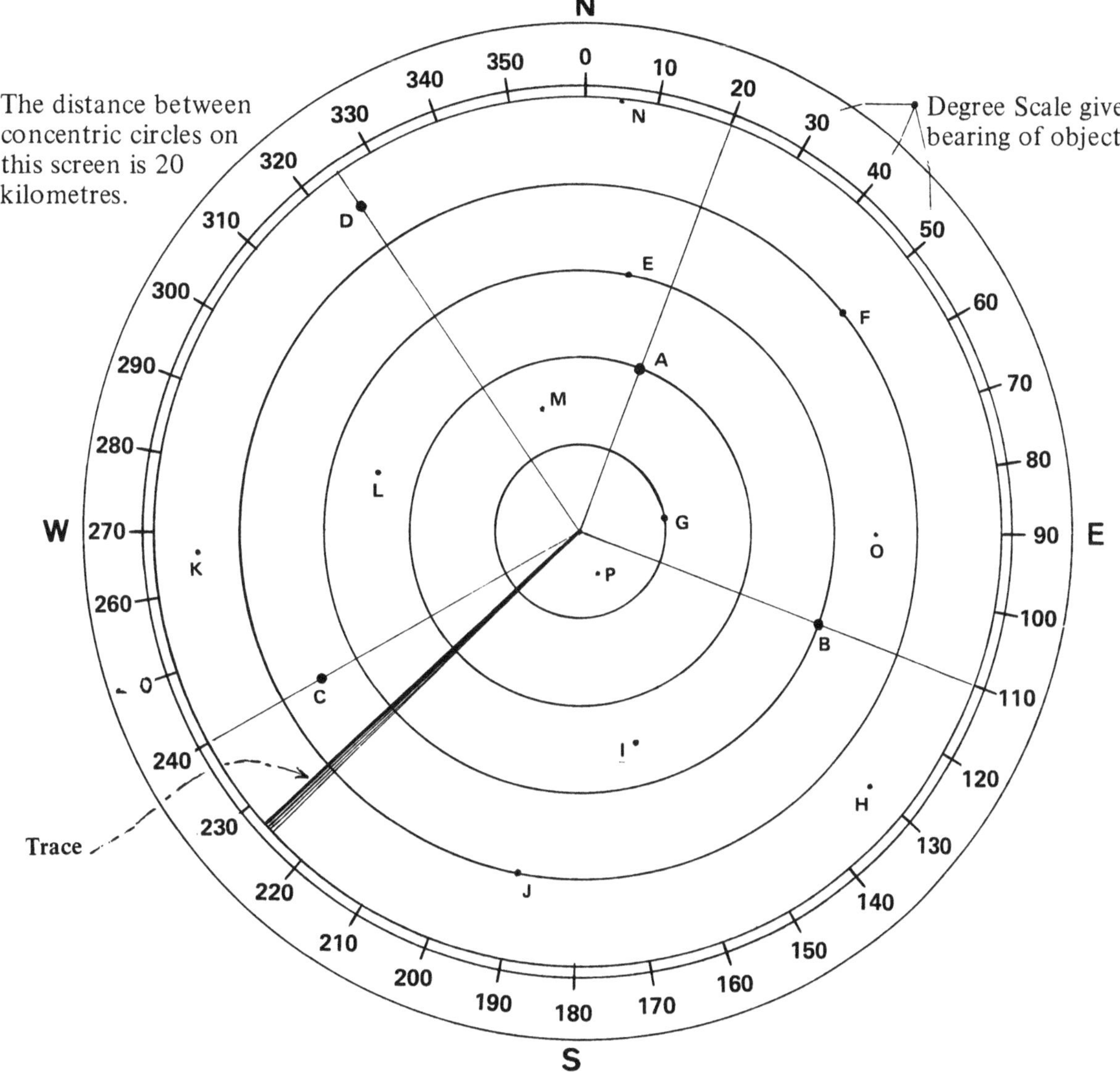

Blips are aqppearing on this screen at positions A to P

A is 40km on a bearing 020°
B is 60km on a bearing 110°

C is 70km on a bearing 240°
D is 90km on a bearing 325°

1. Write the distance and bearing of each of the remaining positions.
Use your Mathomat protractor to help give an accurate bearing.

E is km on a bearing °.
F
G
H
I
J

K
L
M
N
O
P

RADAR

2. Write the distance and bearing of each blip (labelled M to V) that has occured on this radar screen. The distance between concentric circles is 10km ie. scale 1cm = 10km. Draw in a light to represent the "trace" at each blip and use your Mathomat to make accurate measures.

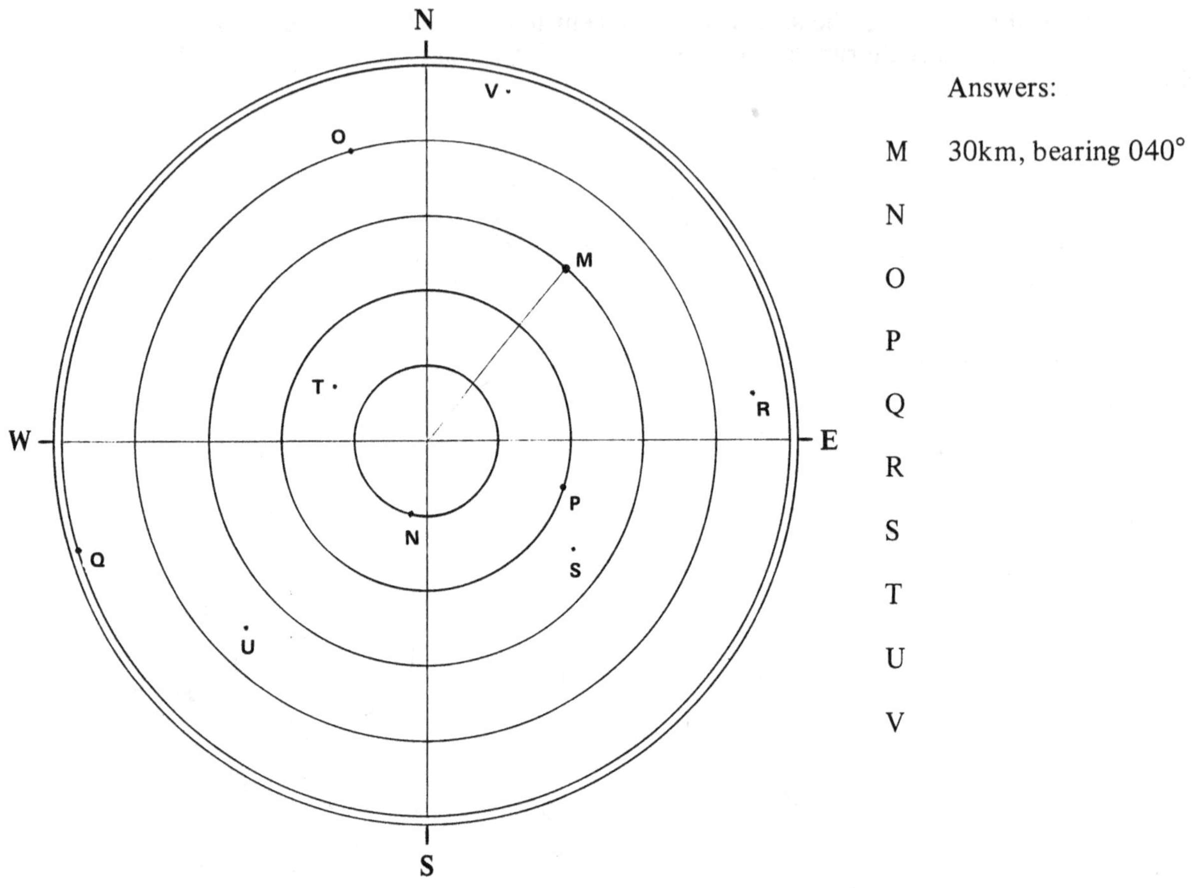

Answers:

M 30km, bearing 040°

N

O

P

Q

R

S

T

U

V

3. Accurately position the following points on the radar screen shown. Scale 1cm = 10km.

A 20km, bearing 015°

B 25km, bearing 035°

C 35km, bearing 090°

D 45km, bearing 160°

E 15km, bearing 175°

F 30km, bearing 208°

G 40km, bearing 247°

H 18km, bearing 265°

I 48km, bearing 310°

J 37km, bearing 340°

K 23km, bearing 355°

L 12km, bearing 027°

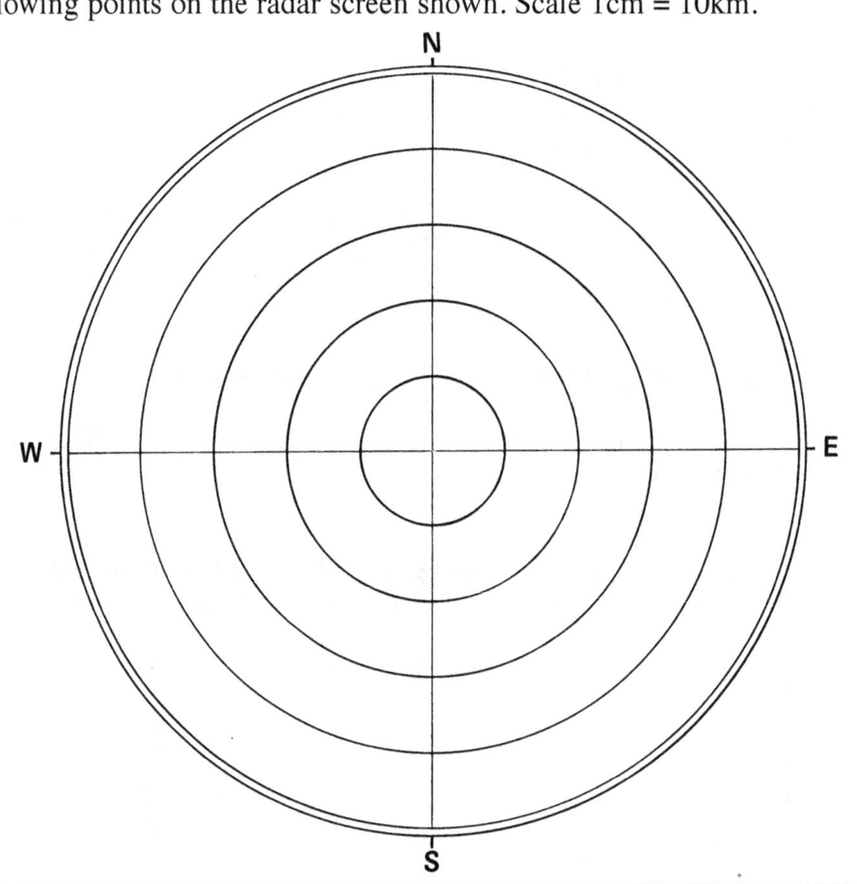

RADAR

4. Draw a radar screen with 5 concentric cicles centred on Point P.
The distance between concentric circles is 10km. Scale 1cm = 10km.

In each case, draw a straight line from P to the dot on the object. Give the istance and bearing of all objects mentioned.

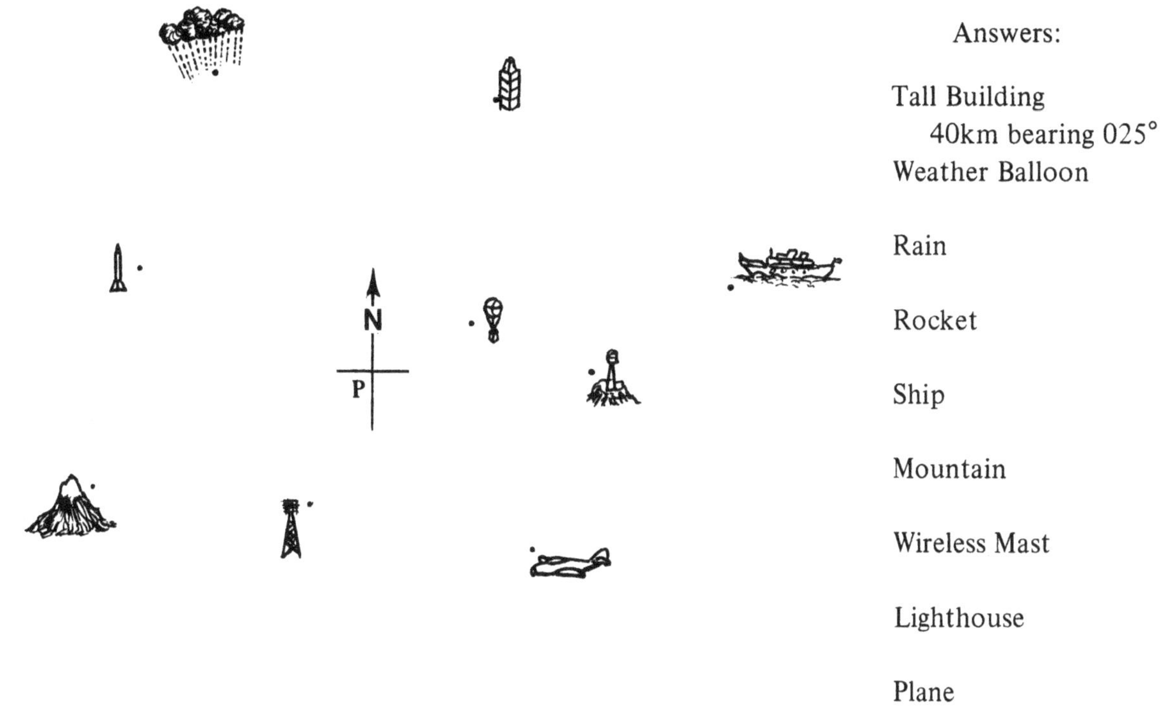

Answers:

Tall Building
40km bearing 025°
Weather Balloon

Rain

Rocket

Ship

Mountain

Wireless Mast

Lighthouse

Plane

5. Draw a radar scren with 5 concentric circles centred on point P.
Th distance between concentric circles is 10km. Scale 1cm = 10km. Now plot these points on the screen.

A. 36km, bearing 023°

B. 29km, bearing 303°

C. 44km, bearing 250°

D. 50km, bearing 080ª

E. 13km, bearing 168°

F. 38km, bearing 204°

G. 22km, bearing 204°

H. 47km, bearing 344°

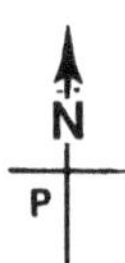

SNOWFLAKE GEOMETRY

1

This Unit of work in Mathomat In struction Text Book Vo. 1 has proved so popular, that some further designs have been included in Vol. 2.

Again, using your Mathomat and the basic hexagonal shape provided complete the patterns shown.

SNOWFLAKE GEOMETRY
2

SNOWFLAKE GEOMETRY

Using the hexagonal shape provided and the sheet of snowflakes for inspiration, see if you can create some further snowflake designs.

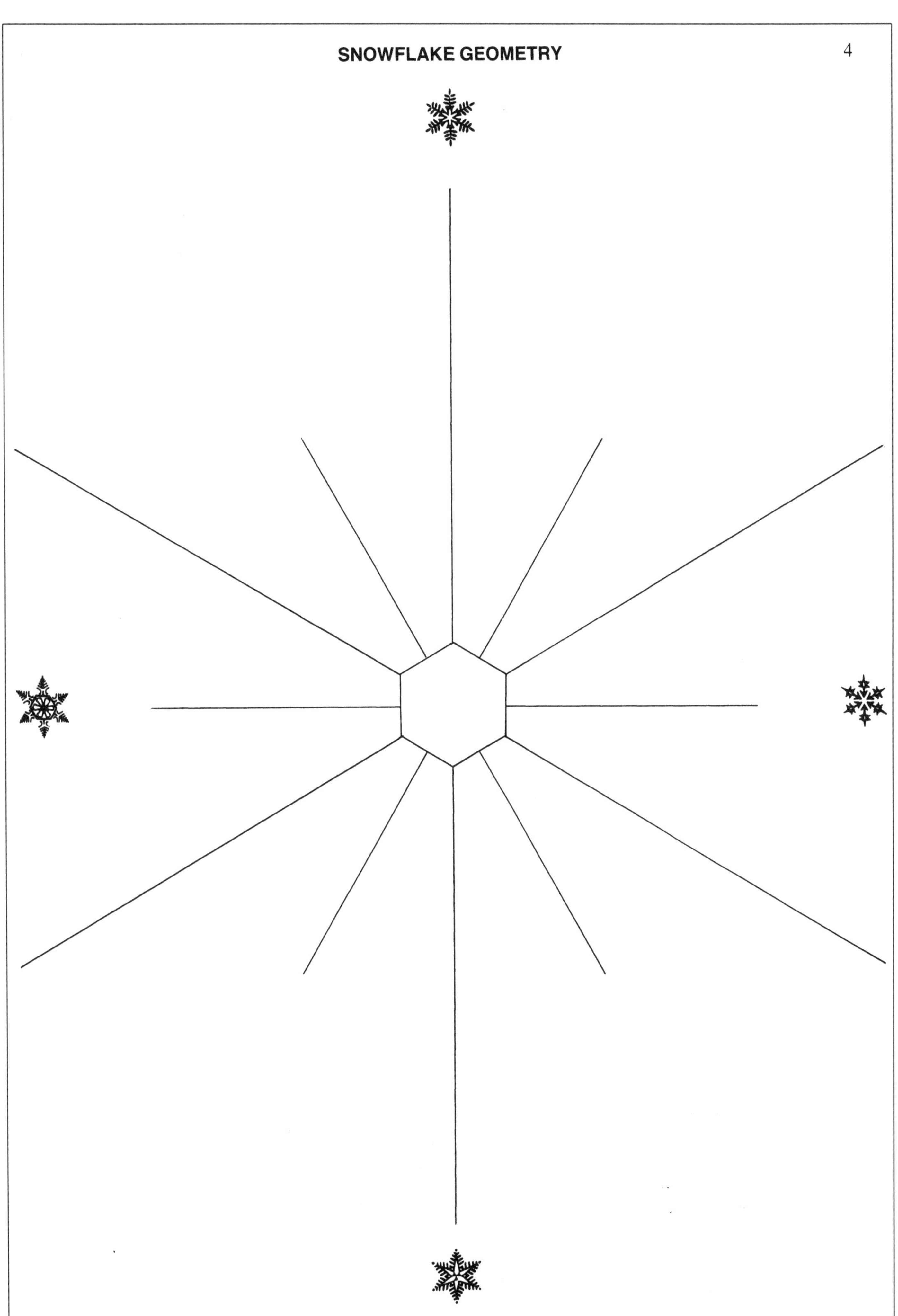
SNOWFLAKE GEOMETRY
4

CONGRUENT FIGURES

Amongst the following figures, there is a pair that are identical in every respect, that is, have the same shape and size. Use your Mathomat to find them.

Write your answer by completing the sentence below.

Answer: Figure ___ is identical in every respect to figure ___

Figures that have the same shape and size are called <u>Congruent Figures</u>.

The sign $\cong$ stands for "is congruent to".

CONGRUENT FIGURES

2

There are twelves different pairs of congruent figures in the following. Use your Mathomat to help you identify these pairs. Write your answers by completing the table below.

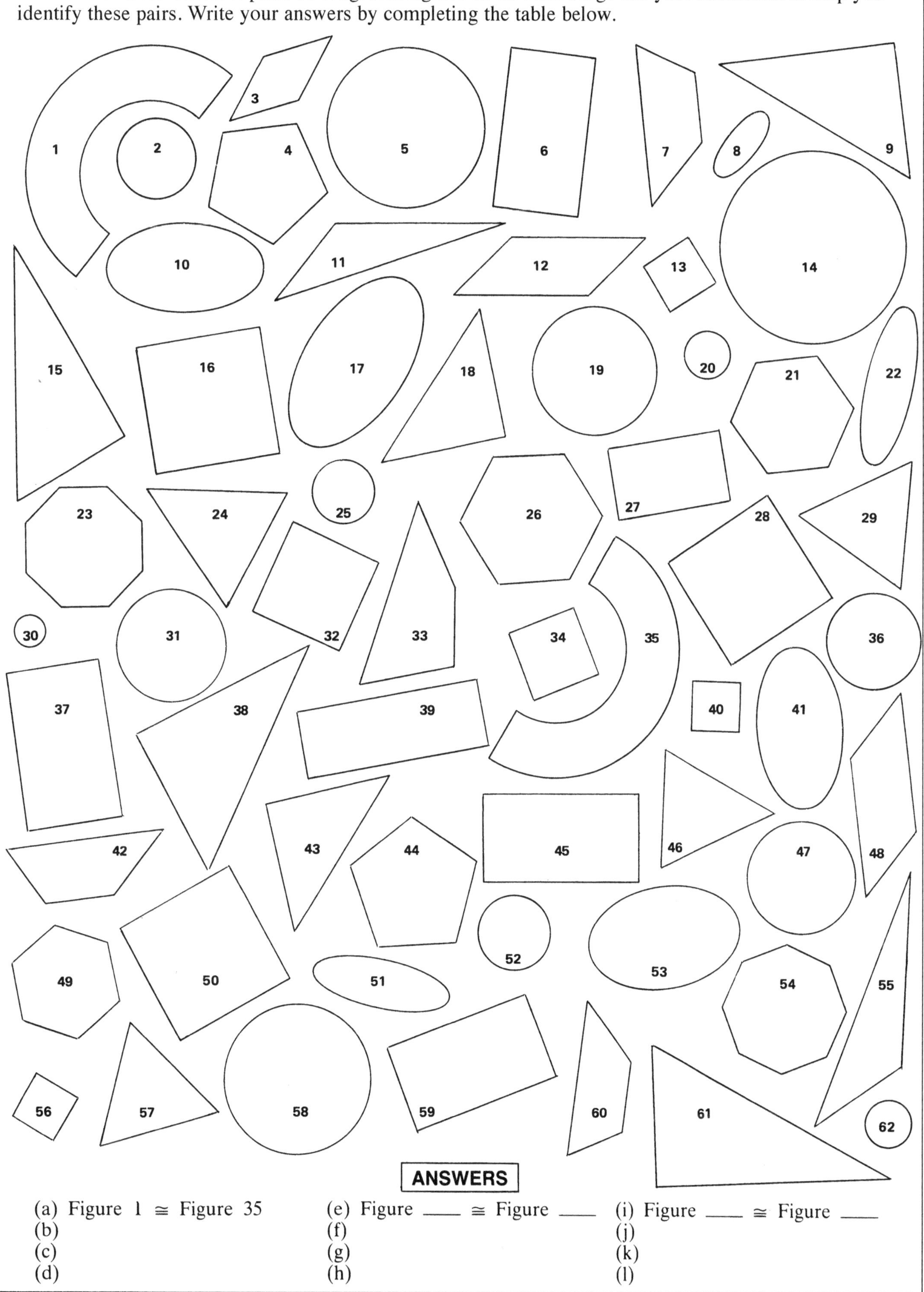

ANSWERS

(a) Figure 1 ≅ Figure 35	(e) Figure ___ ≅ Figure ___	(i) Figure ___ ≅ Figure ___
(b)	(f)	(j)
(c)	(g)	(k)
(d)	(h)	(l)

CONGRUENT FIGURES

Which pair, of the figures shown, is congruent?

(a) 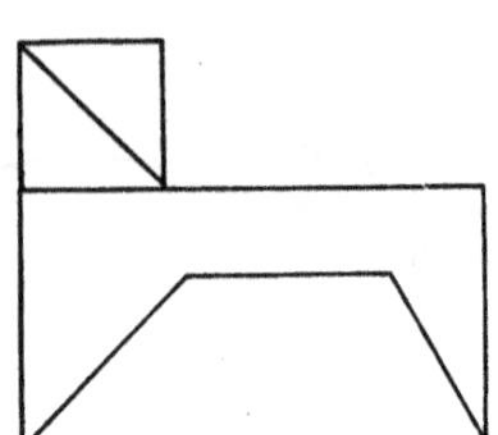

(b) 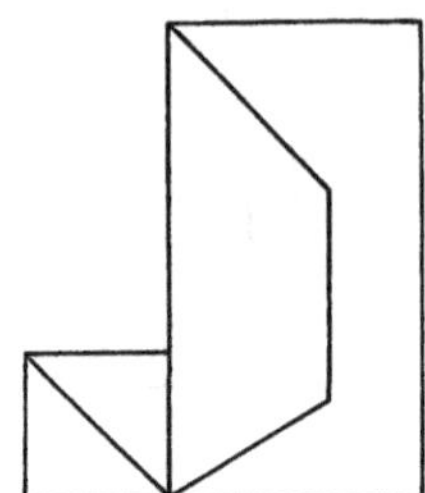

(c) 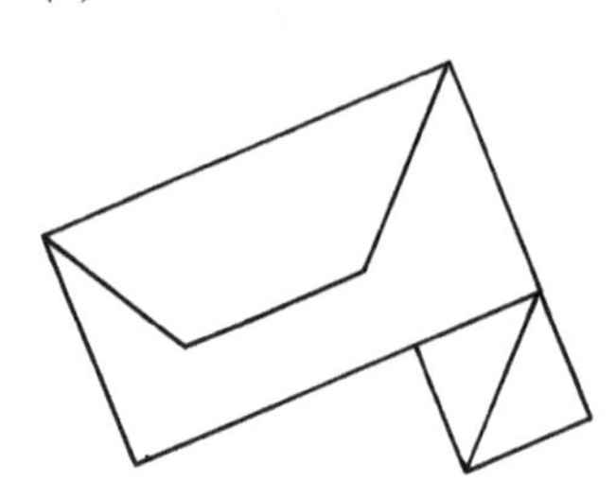

(d) 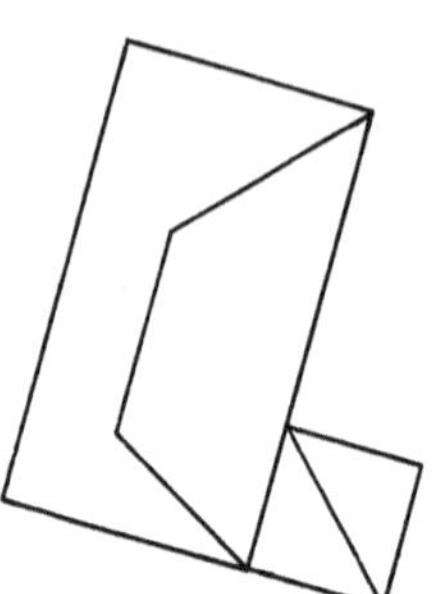

(e) 

(f) 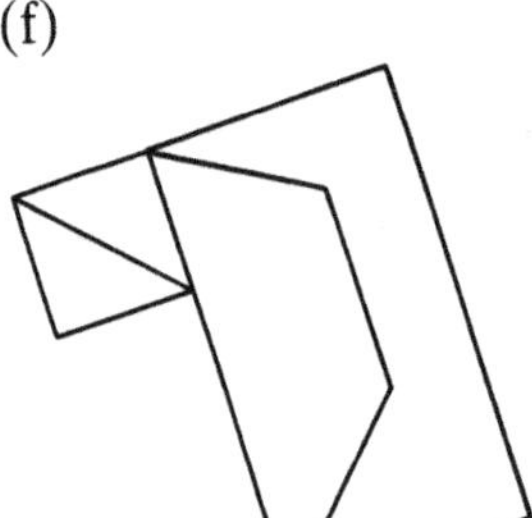

(g) 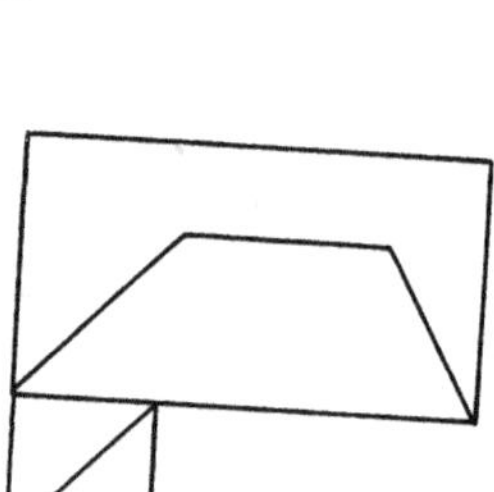

(h) (i) 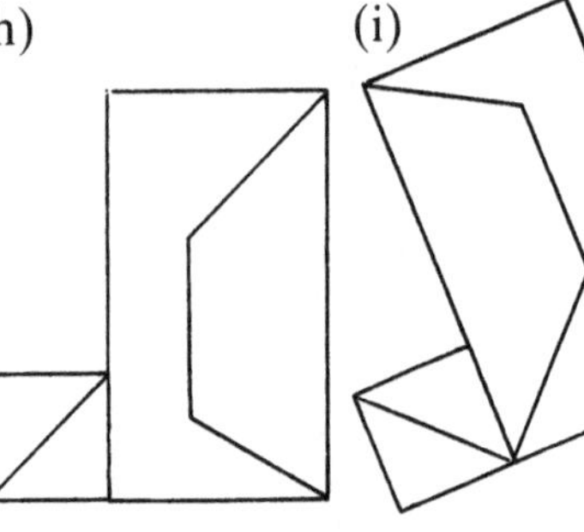

Ans.____________

Which pair, of the three dimensional figures shown, is congruent?

(a) 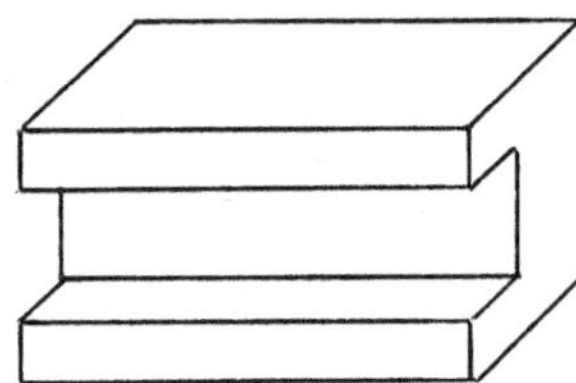

(b) 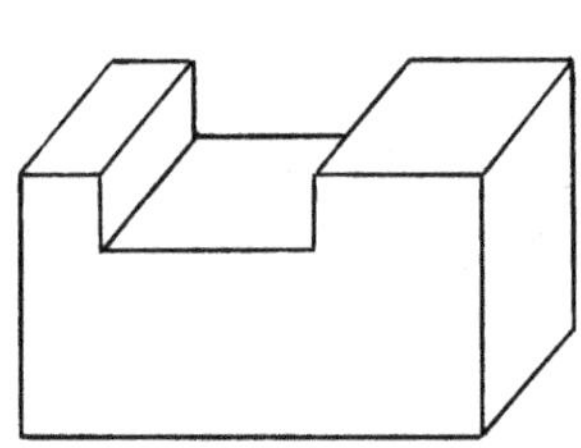

(c) 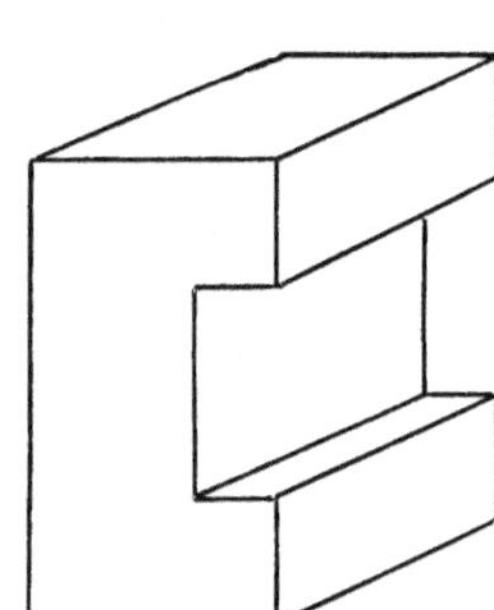

(d) 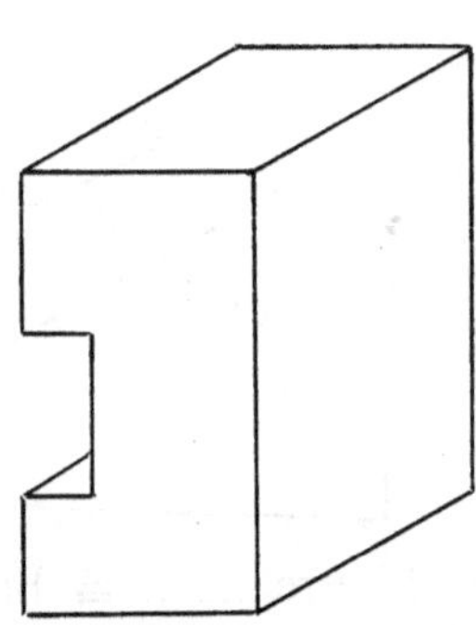

(e) 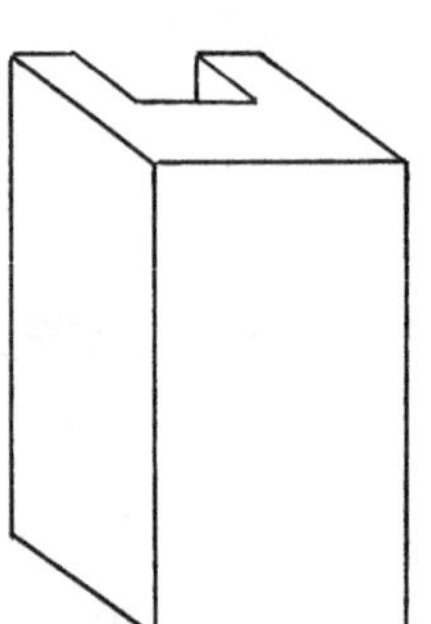

(f) 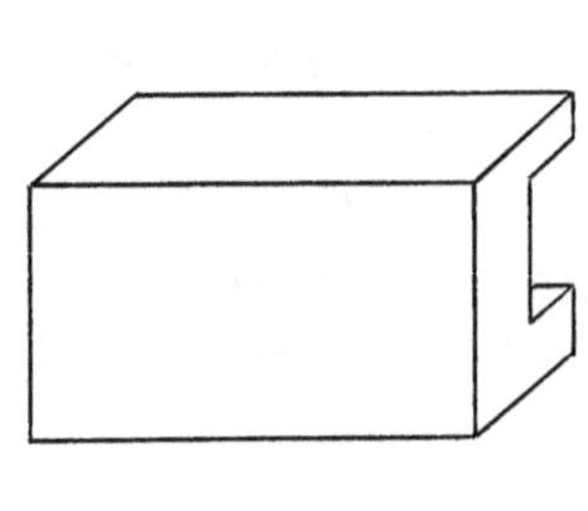

Ans.____________

A pair of figures may be regarded as **congruent** when one igure can be obtained from th other by means of one or more of the following geometrical transformations.

A. Translation

A translation occurs when a figure moves in a particular direction. without turning.

Example

Hexagon ABCDEF has been translated from position 1 to position 2.

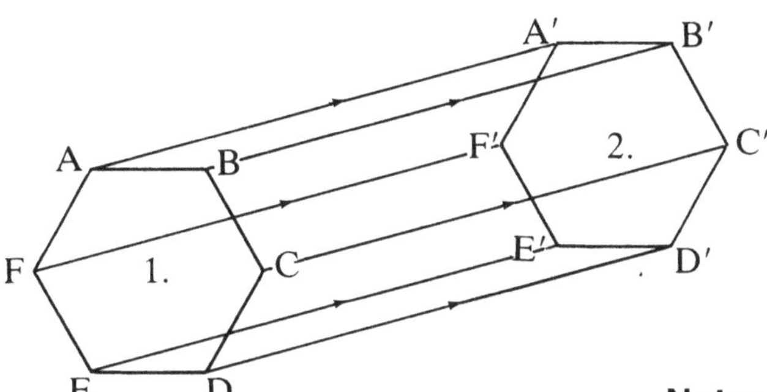

Note: All intervals remain parallel to their originals.

Using translation and your Mathomat complete the following pairs of congruent figures.

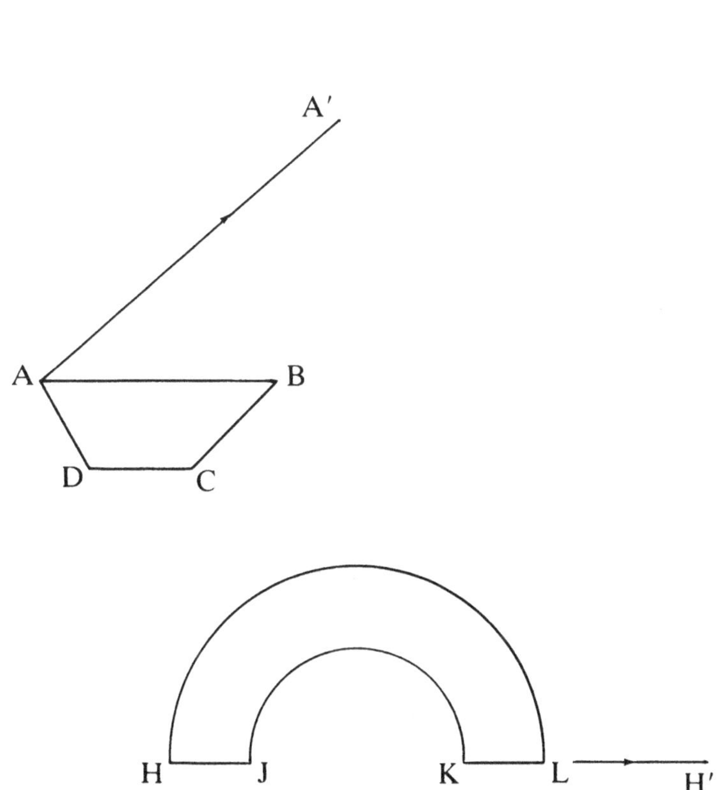

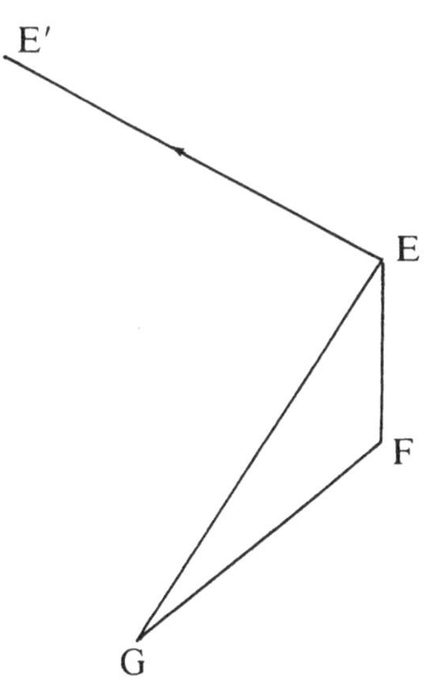

B. Rotation

A rotation occurs when a figure moves in a particular direction about a centre of rotation.

Example

Parallelogram ABCD has been rotated 120° in a clockwise direction about C from position 1 to position 2.

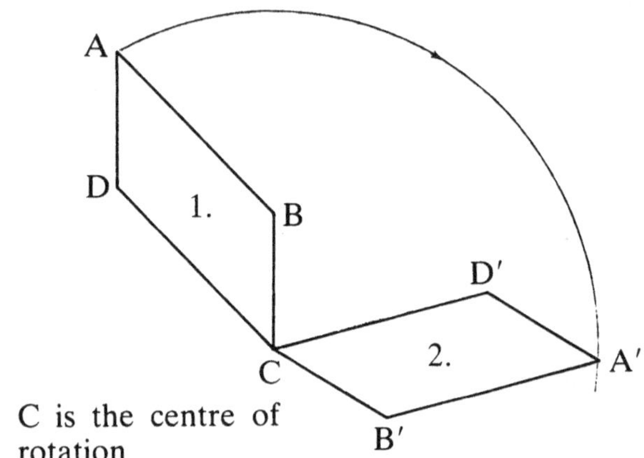

C is the centre of rotation

Steps involved:

(i) Draw Fig. ABCD of your Mathomat.

(ii) Using interval DC measure and draw an angle of 120° in a clockwise directrion.

(iii) Reposition Fig. ABCD and place a biro or pin at C and a pencil at A. Hold steady at C and rotate clockwise to the required position.

Using rotation and your Mathomat complete the following pairs of congruent figures.

(a) Rotate rectangle ABCD 130° in a clockwise direction about D.

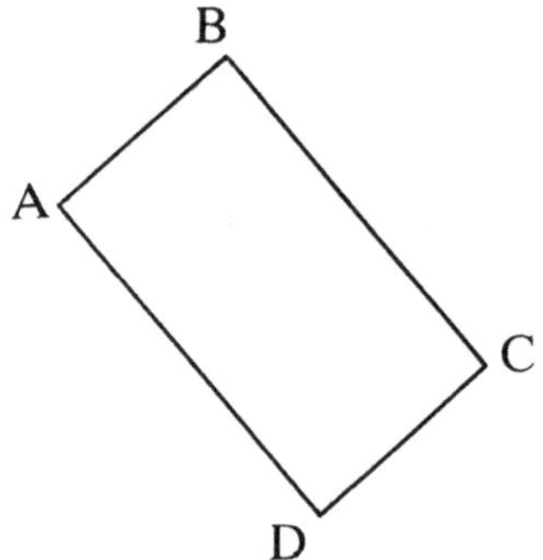

(b) Rotate trinangle EFG 150° in an aqnticlockwise direction about G.

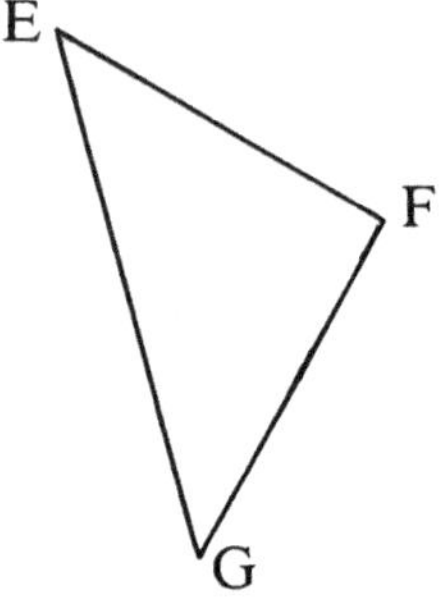

C. Reflection

A reflection occurs when a figure is moved in such-a-way that it becomes'opposite' in sense, that is, left becomes right, clockwise becomes anticlockwise.

Reflction in a mirror is a sense-reversing operation, and like this reflection, the interval taking the place of the miror is clled the mediator or axis of symmetry.

Example

Trapezoid ABCD has been reflected about the mediator 'm' from position 1 to position 2.

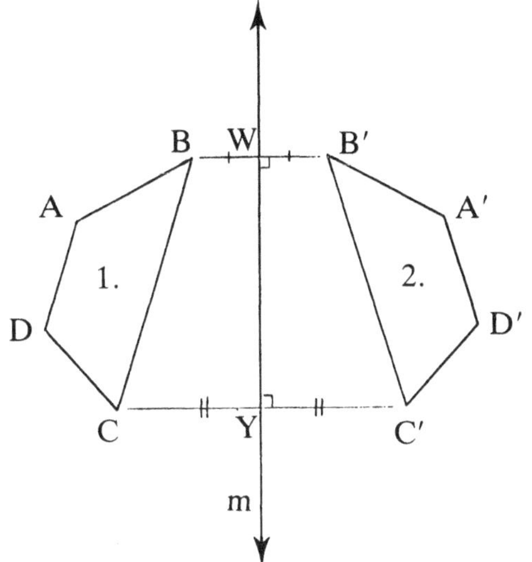

Note:

(i) Your Mathomat had to be turned over in order to obtain the 2nd position.

(ii) The mediator 'm' is the perpendicular bisector of the interval joining a point and its image; That is BW = WB' and CY = YC'.
also BB' and cc' are ⊥ WY

Using reflection and your Mathomat complete the following pairs of congruent figures.

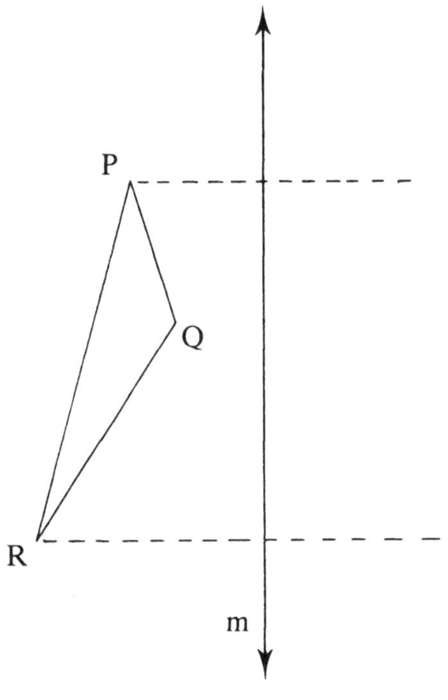

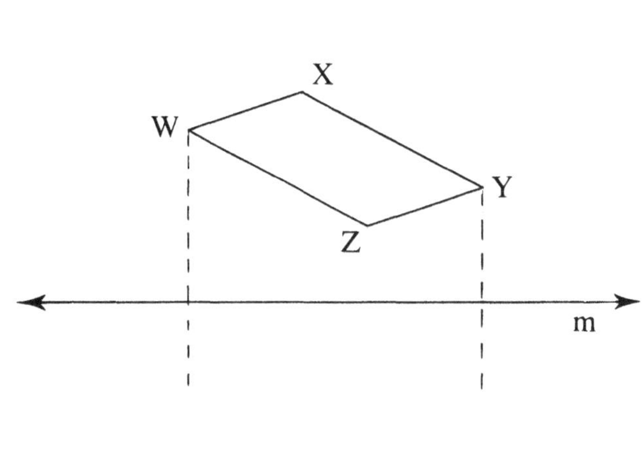

Note:

The figure obtained by one or more of the above transformations is clled "the image" figure.

The set of the above transformations is clled "the set of ISOMETRIC TRANSFORMATIONS". (Isometric means equal measure)

CORRESPONDING PARTS OF FIGURES

Consider the following transformations.

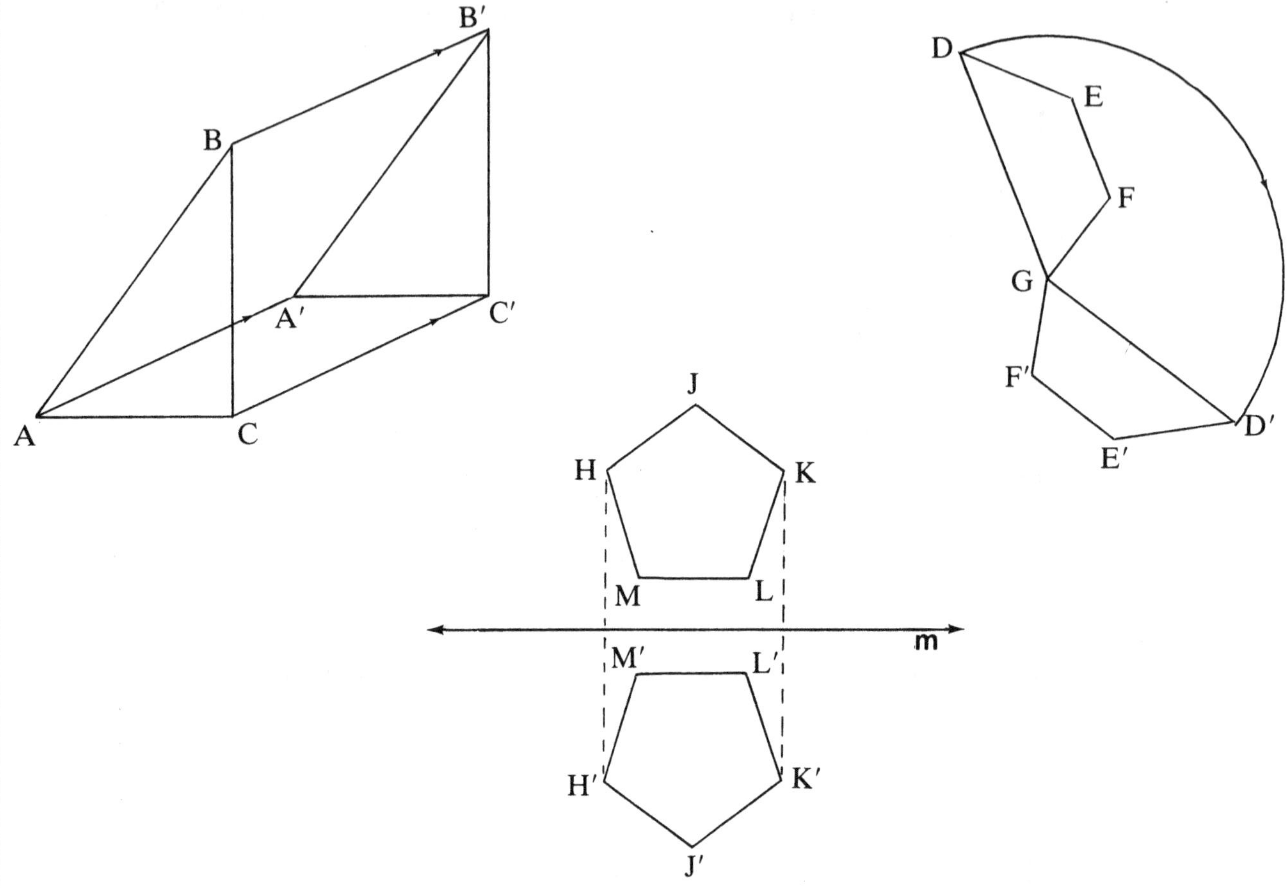

The part of the figures (vertices, intervals, angles) that transform into each other are called **corresponding parts.**

Complete the following

In the above, vertex A corresponds to verted ________
vertex B' corresponds to vertex ________
vertex F' corresponds to vertex ________
vertex J' corresponds to vertex ________
vertex D' corresponds to vertex ________
interval BC corresponds to interval ________
interval E'F' corresponds to interval ________
interval MH corresponds to interval ________
interval GD' corresponds to interval ________
interval AB corresponds to interval ________
angle ABC corresponds to interval ________
angle GD'E' corresponds to interval ________
angle DEF corresponds to interval ________
angle J'K'L' corresponds to interval ________

Congruent Triangles are triangles that are equal in all respects, that is corresponding sides and angles are equal.

However, to decide whether or not triangles are congruent we do not need to know all this information, (that the three angles and the three sides of one triangle are equal in measure to the three angles and the three sides of the other triangle).

Let us now experiment to find the minimum conditions necessary for trinagles to be congruent.

Excercise I

use your Mathomat to complete the following triangle, that is, Fig. 19.

B

A · · C

Given only the lengths of the 3 sides of this triangle, and using a compass, try to construct a triangle that is different to fig. 19 in the space provided. Ine side has already been drawn.

B — C
A — B
A — C

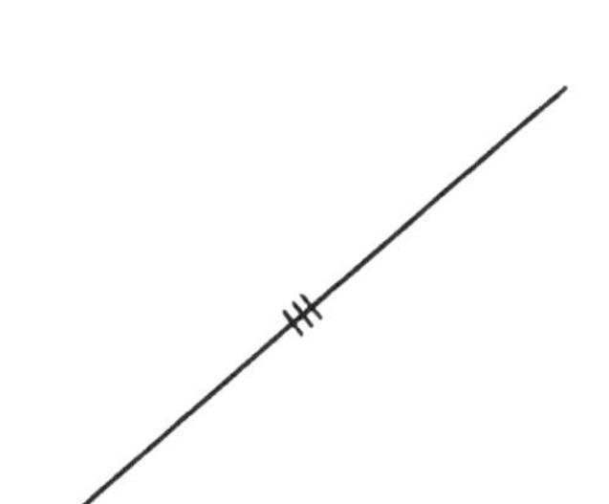

Have you been able to change the shape and size of this triangle?

Ans. ________

1st Condition

Triangles are congruent, if respectively, they have three sides equal (the S.S.S. rule)

Excercise II

Use your Mathomat to complete the following triangle, that is, Fig. 26.

E

D · · F

Given only the lengths of DE and DF and the magnitude of the included angle EDF, try to construct a triangle that is different to Fig. 26 in the spaces provided. Use a compass and protractor, One side has already been drawn.

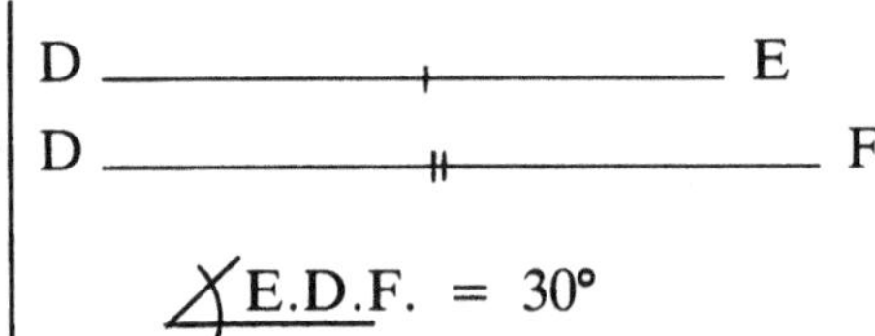

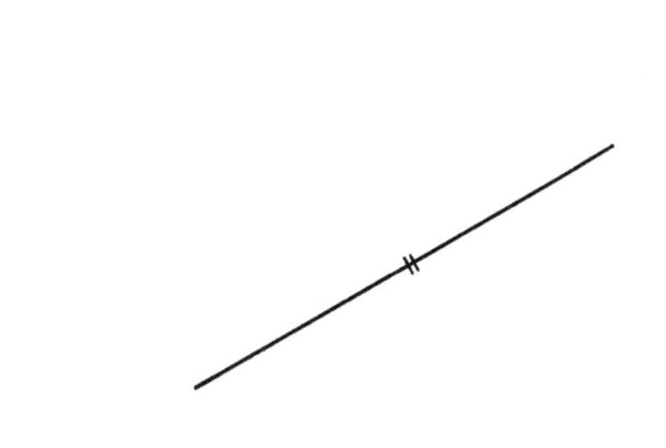

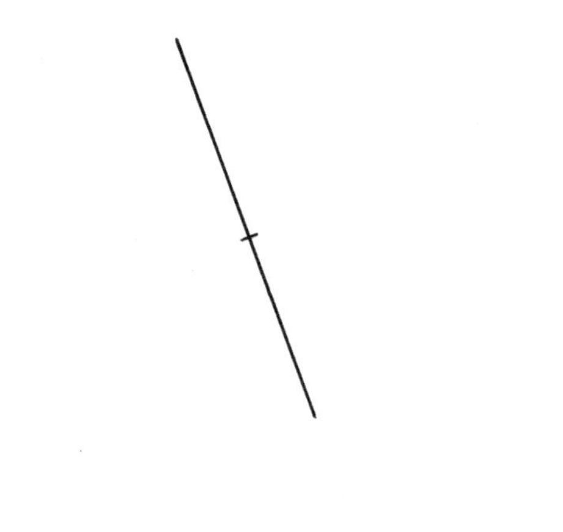

Have you been able to change the shape and size of this triangle?

Ans. ________

2nd Condition

Triangles are congruent, if respectively, they have two sides and the angle formed by those sides equal (the S.A.S. rule).

Exercise III

Use your MATH-O-MATT to complete the following triangle, that is, Fig. 19.

H

G · · J

Given only the lengths of GJ and HJ and the magnitude of a non-included angle HGJ, try to construct a triangle that is different to Fig. 19 in the space provided. Use a protractor and compass. One side has already been drawn.

Hint – Two triangles can be constructed here. See your teacher if you have any difficulty.

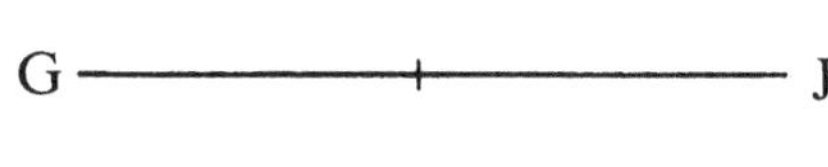

∡HGJ = 18°

Have you been able to change the shape and size of this triangle?

Ans. ____________

Triangle are not necessarily congruent if respectively, they have, two sides and non-included angle equal. S.S.A is not a condition of congruency. This is known as the "ambiguous case" as two triangles of different shape and size can be constructed.

Excercise IV

Use your Mathomat to complete he following triangles, that is Fig. 26.

K · · L

M

Given only the magnitude of angle LKM, the length of KL and the magnitude of angle KLM, try to construct a triangle that is different to Fig. 26 in the spaces provided. Use a protractor. One side has already ben drawn.

∡LKM = 30°

K ———— L

∡KLM = 60°

Have you been able to change the shape and size of this triangle?

Ans. ____________

Again consider triangle KLM.

Given only the magnitudes of angles LKM and KLM and either the length of KM, or LM, try to construct a triangle that is different to Fig. 26 in the space provided. Use a protractor. One side has already been drawn.

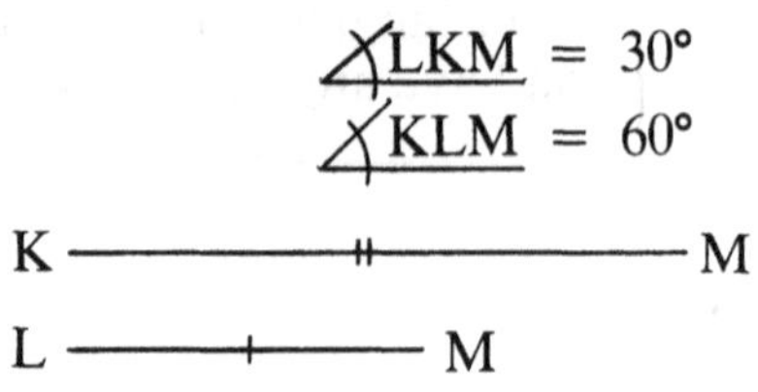

Did you have difficulty in accurately positioning the 2nd angle?

Ans. ____________

Have you been able to change the shape and size of this triangle?

Ans. ____________

3rd Condition

Triangles are congruent, if respectivly, they have two angles and a corresponing side equal (the S.A.S. or A.A.S. rule).

Note:

In the A.A.S. condition, in order to accurately construct this triangle it is necessary to calculate the magnitude of the 3rd angle, so that, in actual fact, we would again have the A.S.A. condition.

Exercise V

Use your Mathomat to complete the following triangles, that is, Fig. 25 (NOP) and its enlargement QRS

CONGRUENT TRIANGLES

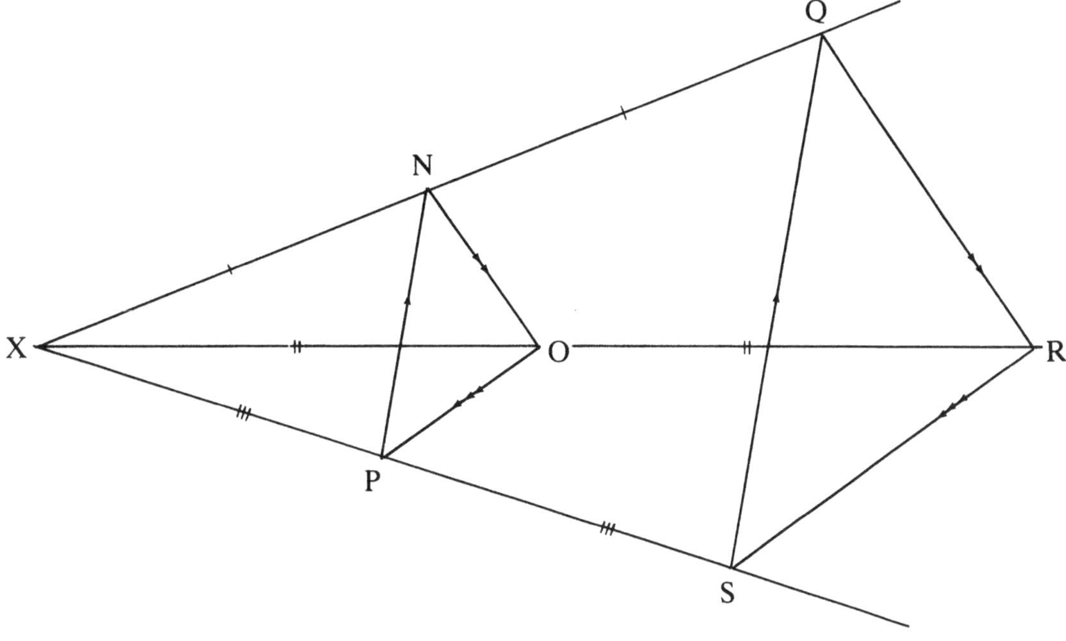

	Answers
Are the corresponding angles of each triangle equal?	________
Have the triangles the same shape?	________
Are the triangles the same size?	________

triangles are not necessarily congruent, if respectively, they have all angles rqual. A.A.A. is not a condition of congruency.

the above triangles NOP and QRS are **similar.**

Exercise VI

Use your Mathomat to complte the following triangle, that is, Fig. 29.

T

V U

Given only the magnitude of angle TUV, the length of the hypotenuse VT, and either the length of TU of VU, try to construct a triangle that is different to Fig. 29 in the spaces provided. Use a protractor and compass. One side has already been drawn.

∡TUV = 90° (a right angle)

V ——————+—————— T

T ——————‖—————— U

V ———‖‖——— U

Have you been able to change the shape and size of this triangle?

Ans.__________

4th Condition

Triangles are congruent, if respectivly, they have a right angle, a hypotenuse, an one other side equal (the R.H.S. rule).

SUMMARY

Triangles are congruent if they have the following corresponding parts equal.

1. S.S.S.

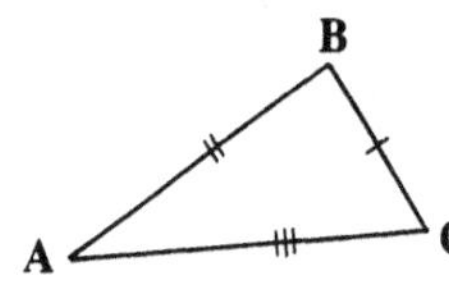

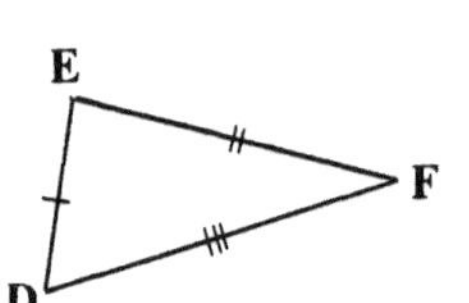

△ ABC ≅ △ DEF

2. S.A.S.

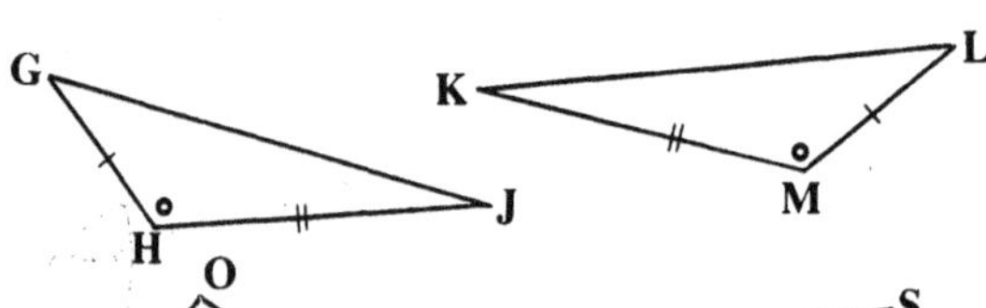

△ GHJ ≅ △ KML

3. A.S.A. or A.A.S.

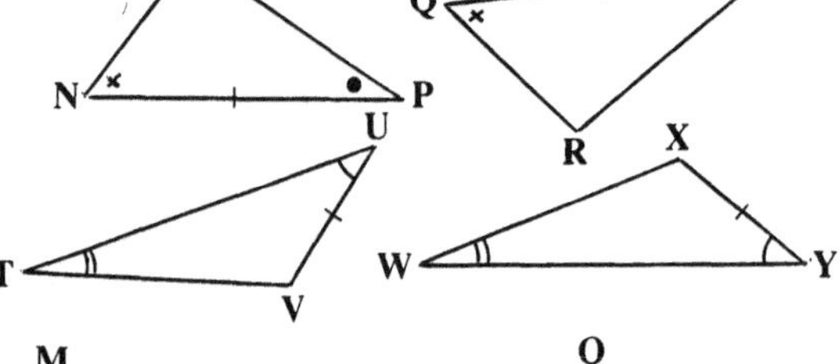

△ NOP ≅ △ QRS

△ TUV ≅ △ WXY

4. R.H.S

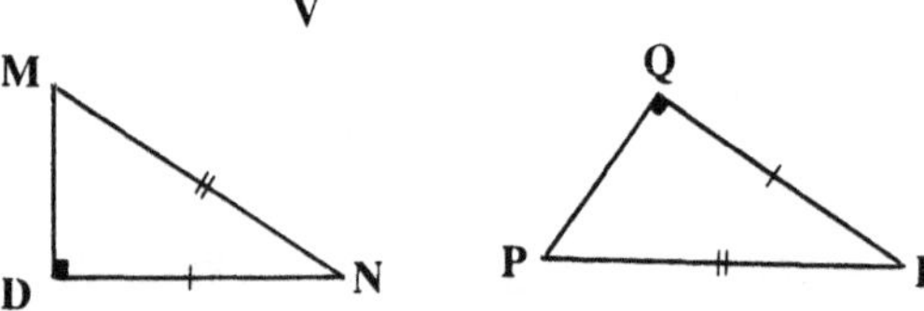

△ MND ≅ △ PQR

Note: A good way of locating whether a side or an angle is a corresponding part, is to see if the angle opposite its side or the side opposite its angle, is the same for the triangles involved.

CONGRUENT TRIANGLES

Examine the followin pairs of triangles, and from the markings on each complete the table shown by deciding whether or not the triangles ar congruent. If congruent, the condition satisfied should be started in abbreviated form.

Triangle Pair	Congruent (Yes or No)	Condition Satisfied	Triangle Pair	Congruent (Yes or No)	Condition Satisfied
1. 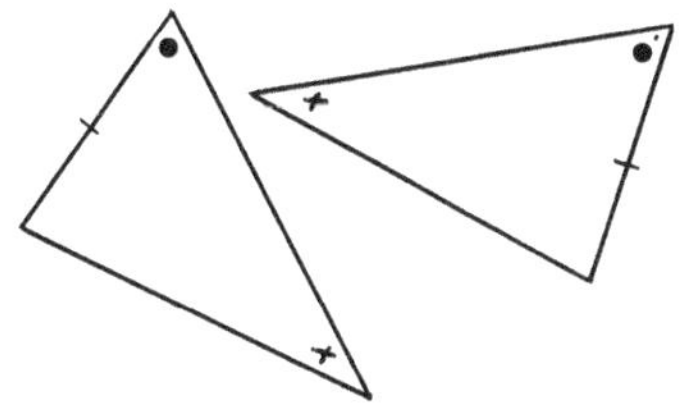	Yes	AAS	7. 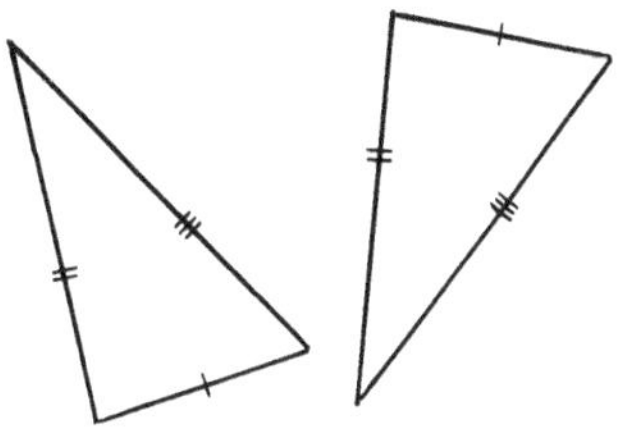		
2. 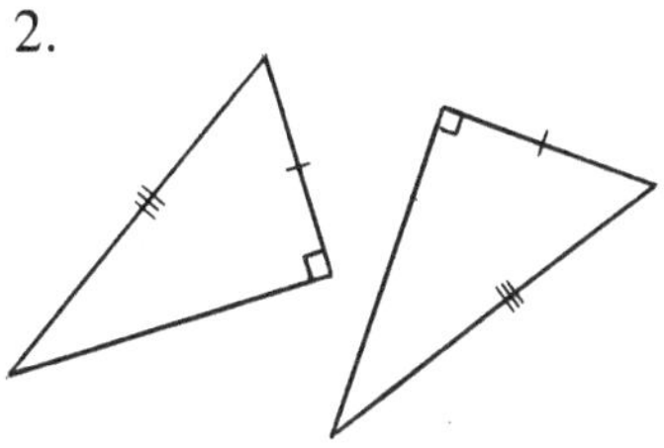			8. 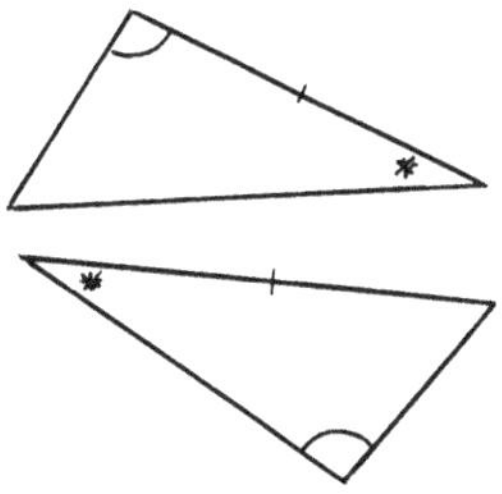		
3. 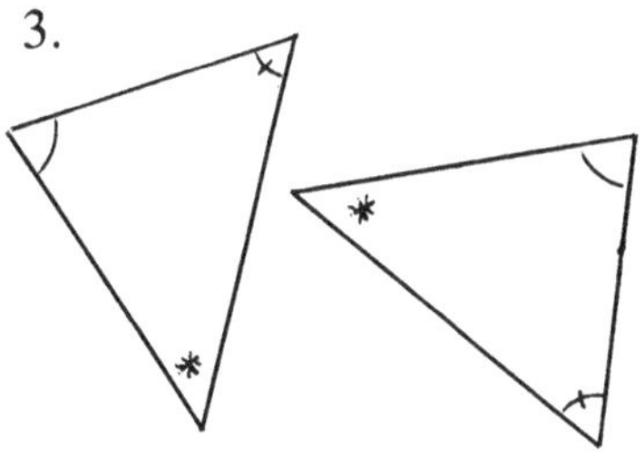			9. 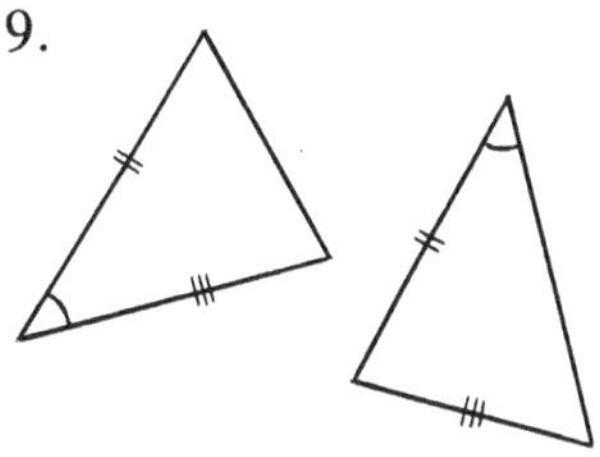		
4. 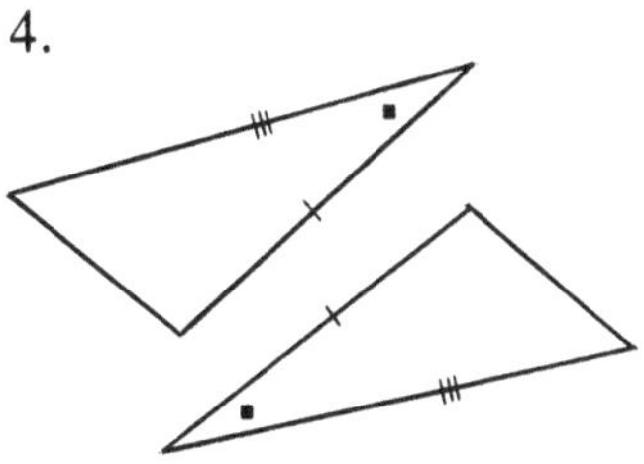			10. 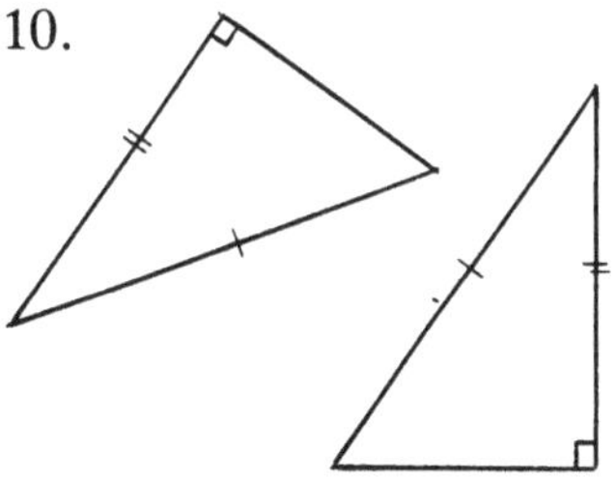		
5. 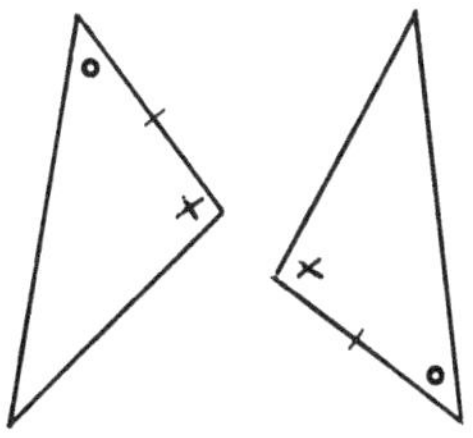			11. 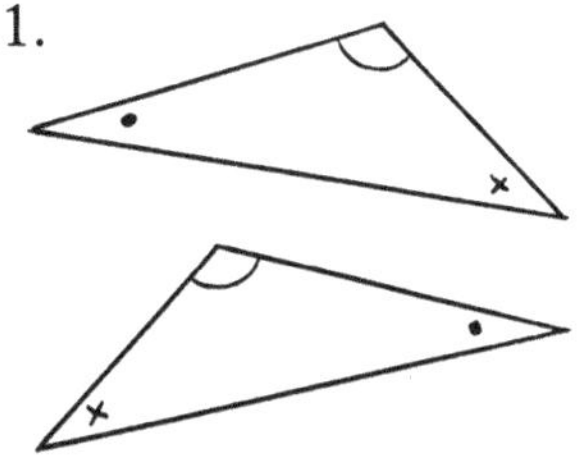		
6. 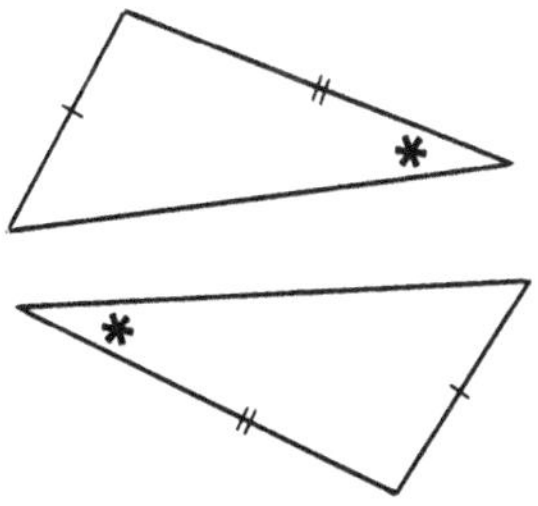			12. 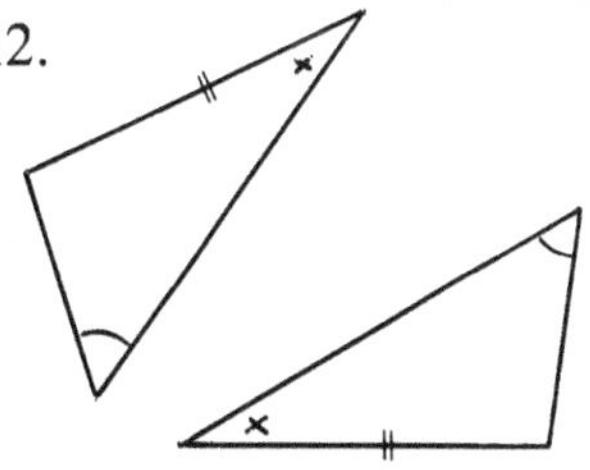		

CONGRUENT TRIANGLES

15

Complete the answer table below by clearly indicating

(i) which triangles are congruent.
and (ii) the condition satisfied. **(Hint)** There are 11 pairs of congruent triangles.

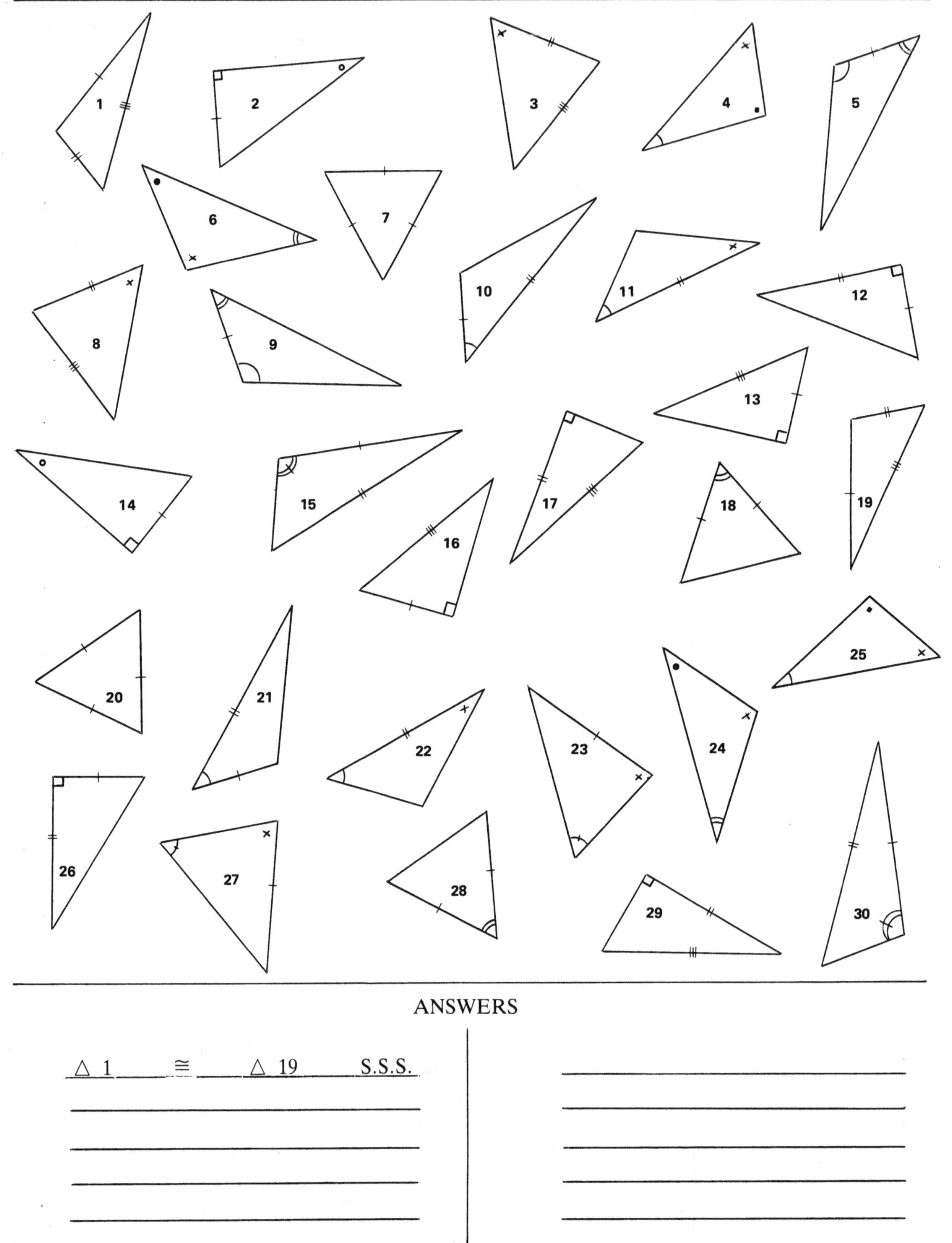

ANSWERS

△ 1 ≅ △ 19 S.S.S.

More Difficult Triangles

Carefully examine the following pairs of triangles and clearly indicate as shown

(i) which triangles are congruent or not.

and (ii) the condition satisfied.

1. 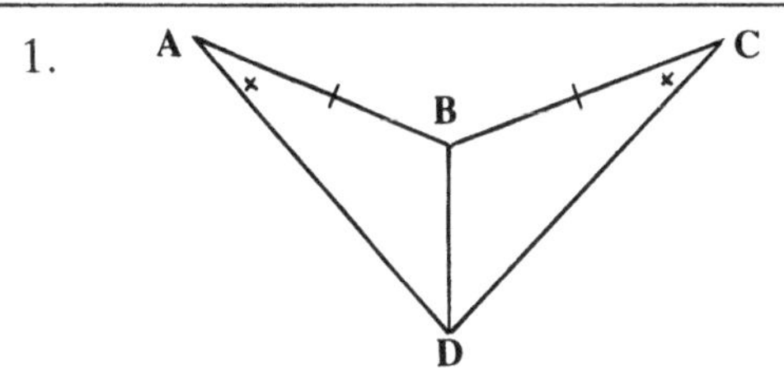	BD is a common side △ ABD ≢ △ CBD (S.S.A. is not a condition of congruency) ≢ means 'is not congruent to'.
2. 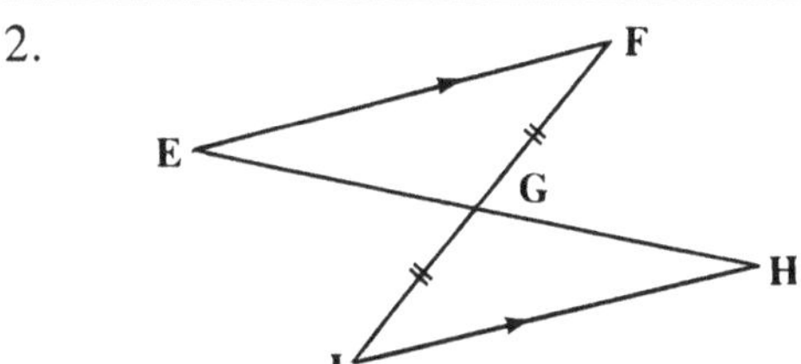	
3. 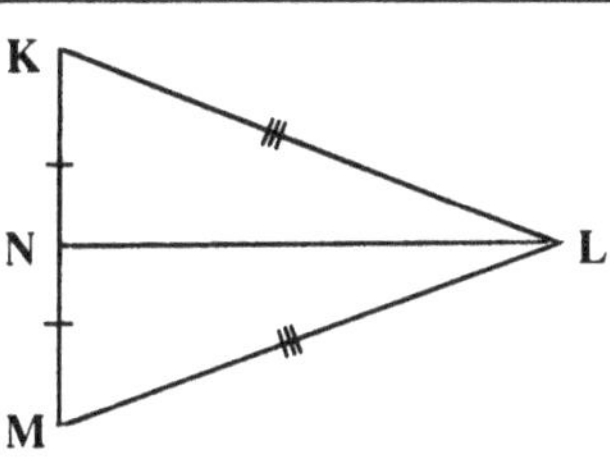	
4. 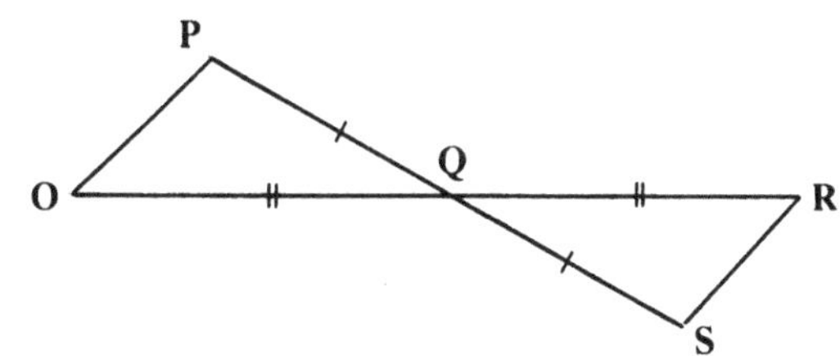	
5. 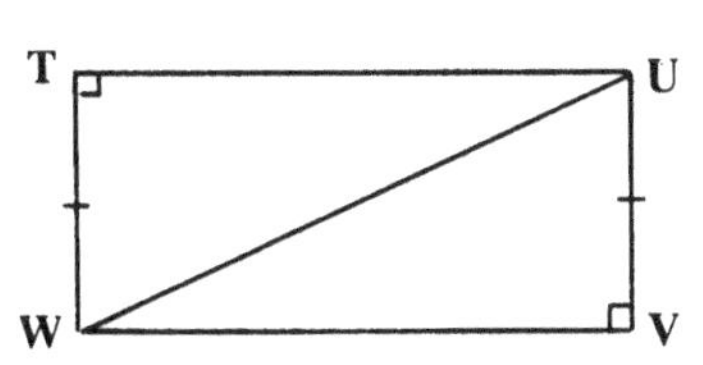	
6. 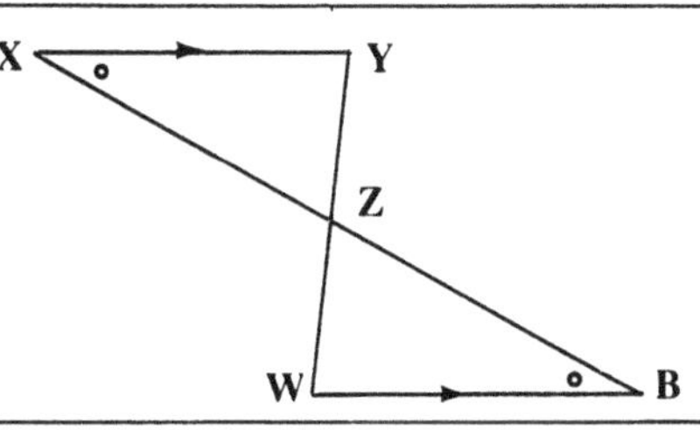	
7. 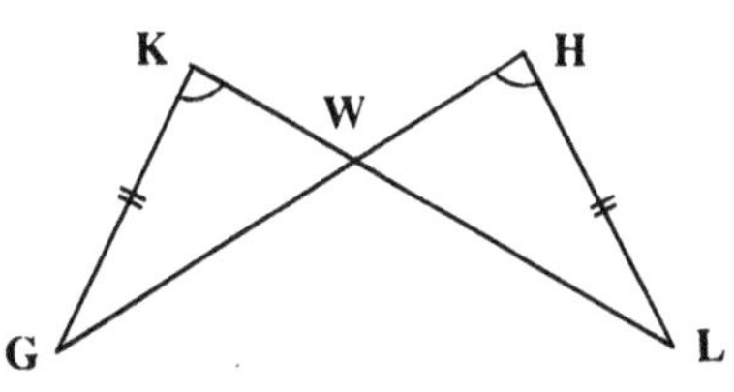	

Mathomat Plastic Cut-outs:

Considerable enjoyment and satisfaction can be obtained by both students and teachers in arranging Mathomat plastic cut-outs into geometric designs, shapes, figures, patterns etc. on sheets of tempered masonite and then "fixing" thses designs with a quick drying spray paint.

the equipment involved consists of

(i) varios sized sheets of tempered masonite.
(ii) a box of plastic cut-outs
(iii) quick drying spray can paint (Coper, Gold, Silver, Black, Clear)

Procedure:

1. Spray the smooth side of the masonite sheet with either coper, gold or silver paint depending on the background colour chosen for th particular design.

2. Plan your design, use light guide lines if neccessary.

3. Place the masonite sheet flatt on a floor or bench.

4. Carefully plase the plastic cut-outs in position.

5. Take a cn of black paint and shake well.

6. Hold the can in such a way that the spray mists down onto the design. (Too close will blow the pieces off the design) Do this lightly the first time and allow approximately 2 to 3 minutes to set.

7. Finish the spraying and wait until the paint has just become tacky.

8. Quickly lift off each individual piece of plastic with the fore-finger. The tackiness of the paint will allow the pieces to stick to your finger and be easily removed to a waste bin.

9. Allow the finished design to completely dry.

10. Spray the design with clear lacquer to, in particular, prevent finger marking of the metallic paints used.

The above steps should be tired on a scrap piece of masonite first to discover the subtleties and techniques involved before trying a complicated design.

A Box of plastic cut-outs can be obtained from

EDUCATION EFFICIENCY PRODUCTS
3 Glencairn Avenue
Ringwood, Vic. 3134
AUSTRALIA. Phone: 870 1738

Each box contains approximately 500 separate pieces of each geometricqal shape in the Mathomat template.

Alternatively to the above, designs may be formed by painting the plastic cut-outs in various colours, then glueing them into position.

Designs for place-mats, coffee tables, decortive poster, wall pictures etc. can be very quickly completed using the techniques mentioned.

* * * * *

GEOMETRICAL CREATIVITY

The following sheets contain ideas that may be useful in creating your own designs.

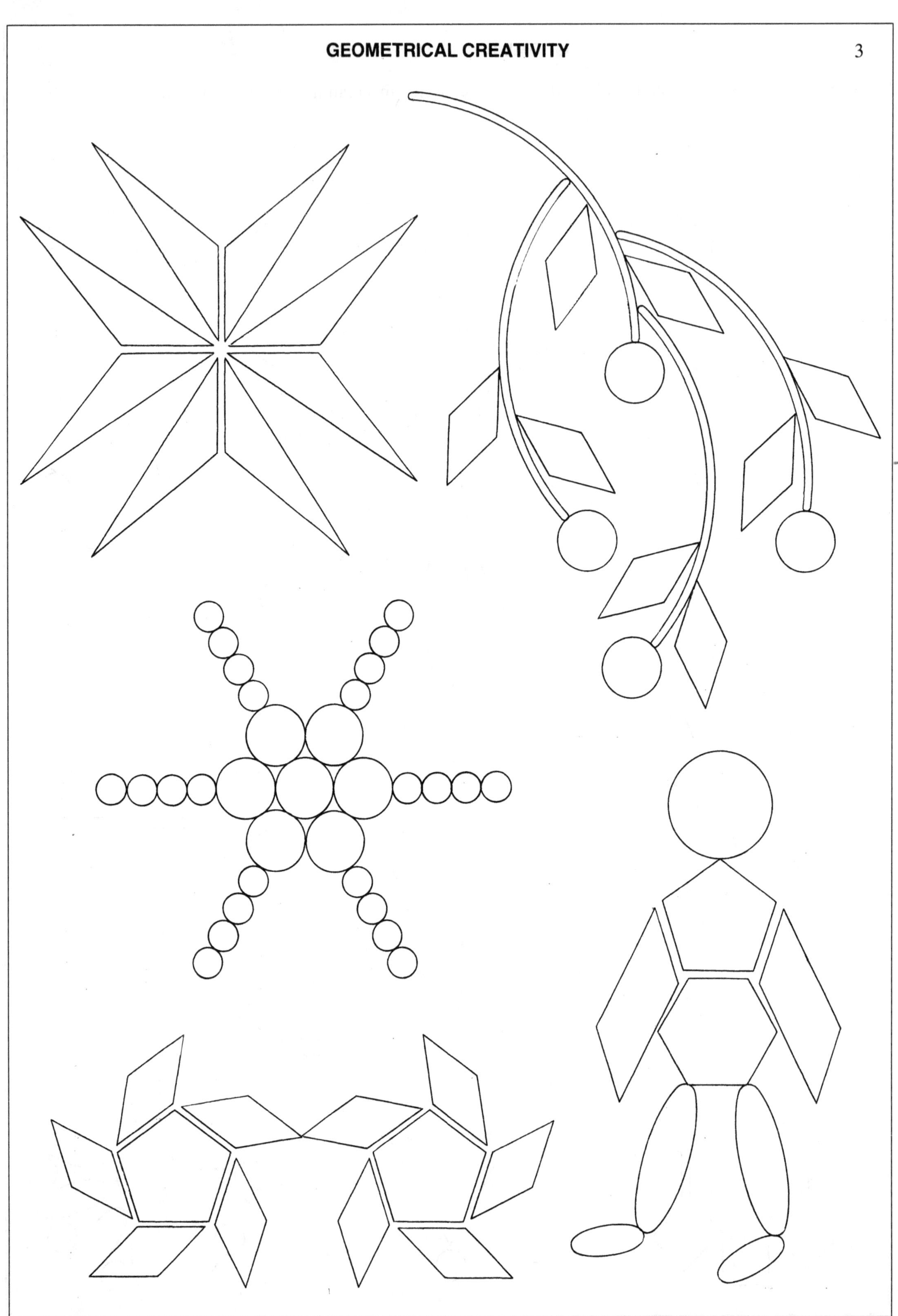
GEOMETRICAL CREATIVITY
3

GEOMETRICAL CREATIVITY
4

RADIAN MEASURE

Look closely at the scales on your Mathomat protractor. You have already measured angles in degrees and in so doing wondered what the other scale was on the outside of the protractor. It is a scale for measuring angles in radians. A radian is simply a bigger and often more convenient unit to use than one degree.

$$1 \text{ radian} = 57.29578°$$
$$\text{or } 1 \text{ rad} \simeq 57.3°$$

> A radian is the angle subtended at the centre of a circle by an arc equal in length to the radius (r)

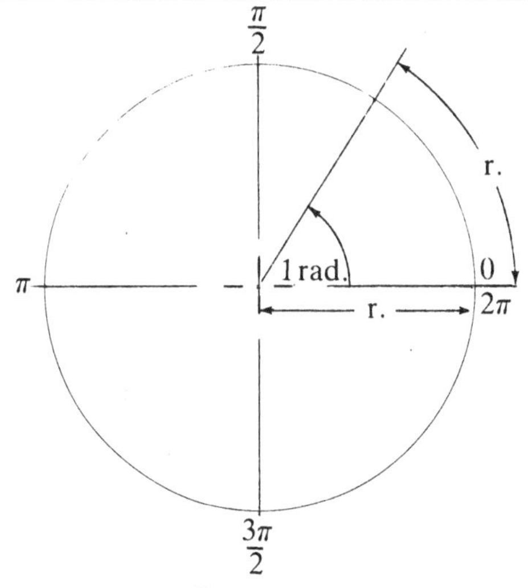

The arc length around the whole circle is equal to the measure of the circumference (**C**)

ie: **C** $= 2\pi r$

The number of radians subtended at the centre of the circle $= \frac{2\pi r}{r}$

$= 2\pi$

⇒ 2π radians $= 360°$

⇒ **π radians $= 180°$**

⇒ 1 radian $= \frac{180°}{\pi}$

$= \frac{180°}{3.1416}$

$= 57.29578°$

$\simeq 57.3°$

also ⇒ 1 degree $= \frac{\pi}{180}$ rad.

1. Complete by marking in the angles on this circle in both radians and degrees.

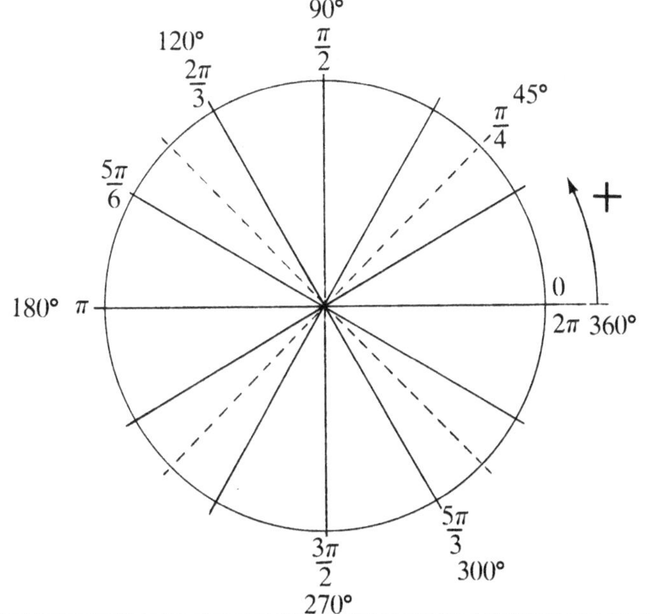

2. Complete the following table.

π	=	3.1416	=	180°
2π	=		=	
$\frac{\pi}{2}$	=		=	
$\frac{3\pi}{2}$	=		=	

Remember these angles.

RADIAN MEASURE

2

Now use your Mathomat to measure the following angles in radians.

1.

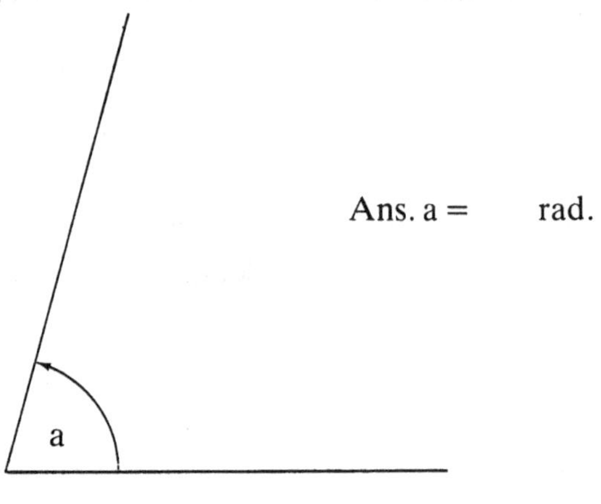

Ans. a = rad.

2.

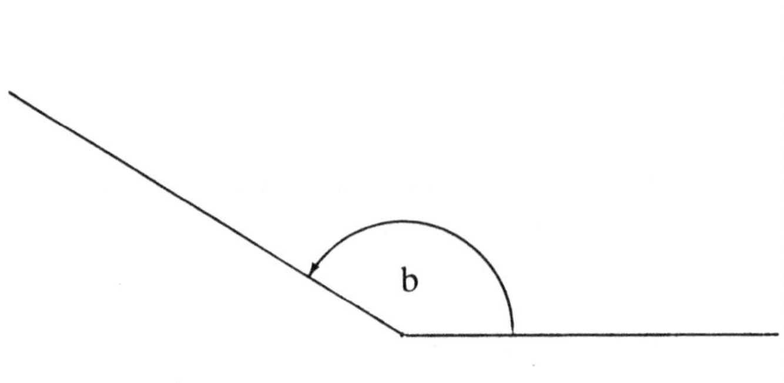

Ans. b = rad.

3.

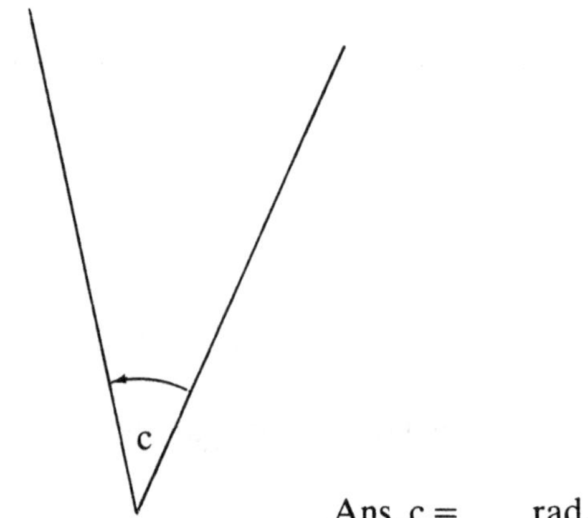

Ans. c = rad.

4.

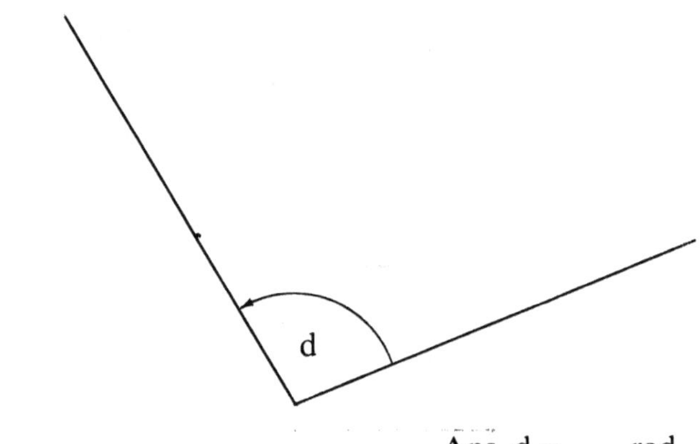

Ans. d = rad.

5.

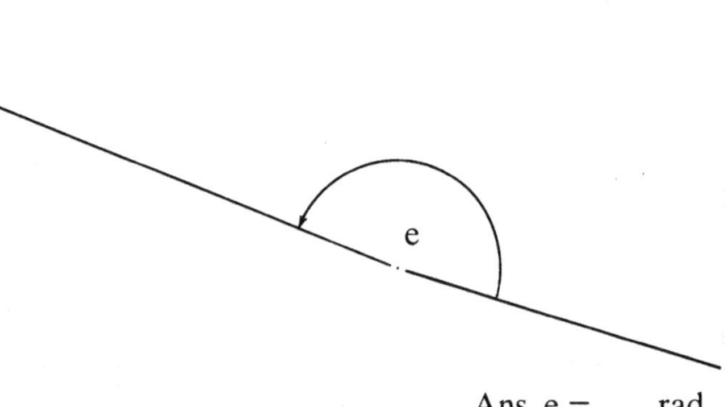

Ans. e = rad.

6.

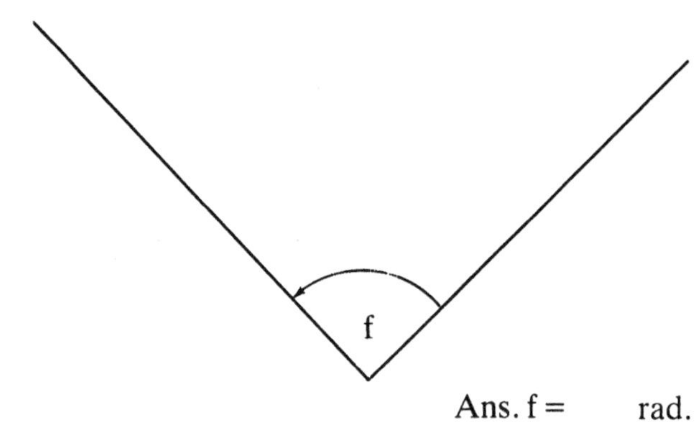

Ans. f = rad.

7.

Use $2\pi = 6.28$

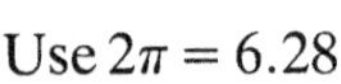

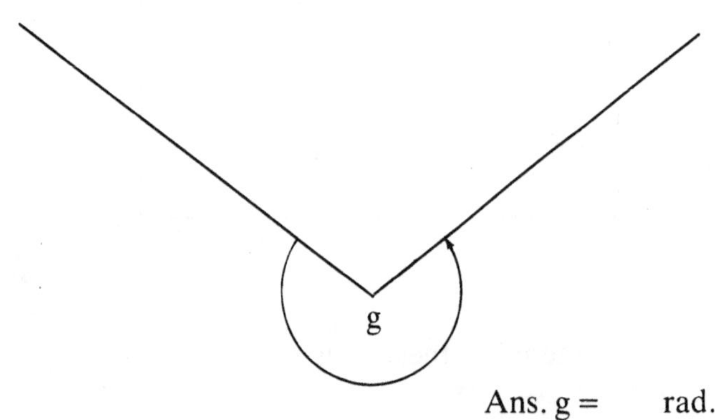

Ans. g = rad.

8.

Use $2\pi = 6.28$

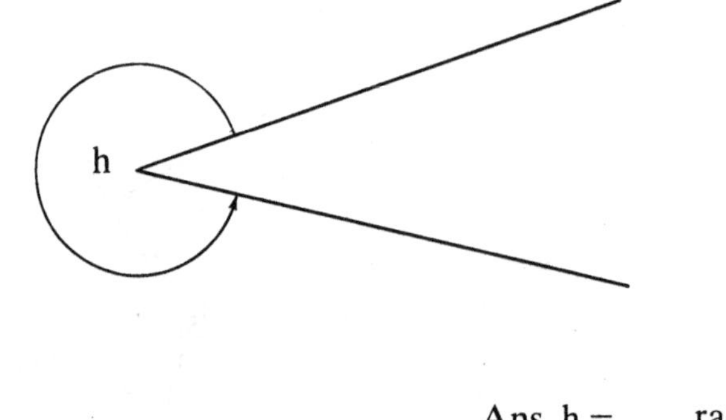

Ans. h = rad.

RADIAN MEASURE

3

Use your Mathomat to draw the following angles. In each case the initial arm of the angle has been drawn. Clearly indicate the angle.

1. 0.95 rad.	2. 2.05 rad.
3. 1.375 rad.	4. 3.05 rad.
5. 1.05 rad.	6. 4.93 rad. (Use $2\pi = 6.28$)
7. 3.75 rad. (Use $2\pi = 6.28$)	8. 5.63 rad. (Use $2\pi = 6.28$)

RADIAN MEASURE

To change radians to degrees. $\left(\times \frac{180}{\pi}\right)$

1. Express in degrees the angles with the following radian measures.

(a)	$\frac{\pi}{6}$	$= \left(\frac{\pi}{6} \times \frac{180}{\pi}\right)^\circ$	$= 30^\circ$	(f)	$\frac{5\pi}{6}$	=		=
(b)	π	=	=	(g)	$\frac{3\pi}{8}$	=		=
(c)	$\frac{\pi}{3}$	=	=	(h)	$\frac{13\pi}{10}$	=		=
(d)	$\frac{3\pi}{4}$	=	=	(i)	$\frac{5\pi}{12}$	=		=
(e)	$\frac{\pi}{2}$	=	=	(j)	$\frac{11\pi}{4}$	=		=

To change degrees to radians. $\left(\times \frac{\pi}{180}\right)$

2. Express the following angles in radians in terms of π

(a)	45°	$= \left(45 \times \frac{\pi}{180}\right)$ rad.	$= \frac{\pi}{4}$ rad.	(f)	315°	=		=
(b)	270°	=	=	(g)	18°	=		=
(c)	72°	=	=	(h)	540°	=		=
(d)	225°	=	=	(i)	200°	=		=
(e)	20°	=	=	(j)	105°	=		=

3. Using tables express the following angles in radians.

(a)	35°	= 0.6109 rad.	(f)	29° 39′	=	(k)	17° 58′	=
(b)	83°	=	(g)	87° 15′	=	(l)	2° 43′	=
(c)	5° 18′	=	(h)	9° 21′	=	(m)	39° 9′	=
(d)	75° 30′	=	(i)	44° 5′	=	(n)	66° 41′	=
(e)	10° 54′	=	(j)	57° 18′	=	(o)	89° 59′	=

4. Using tables and the radian and degree angles given in the diagram, express the following angles in radians.

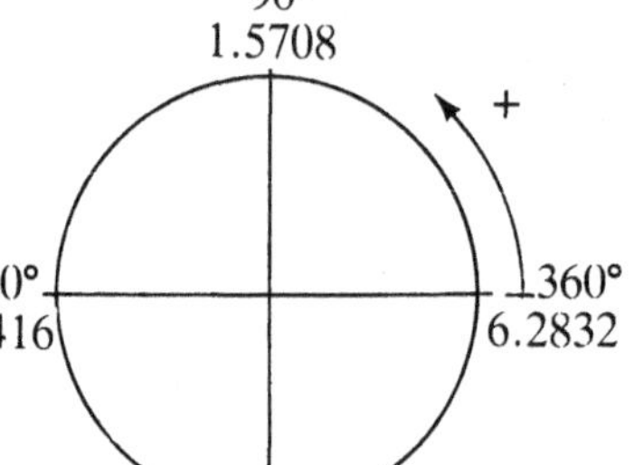

(a)	130°	= 90° + 40°	= 1.5708 + 0.6981	= 2.2689 rad.
(b)	157°	=	=	=
(c)	220°	=	=	=
(d)	235°	=	=	=
(e)	265°	=	=	=
(f)	119° 16′	= 90° + 29° 16′	= 1.5708 + 0.5108	= 2.0816 rad.
(g)	156° 30′	=	=	=
(h)	200° 27′	=	=	=

(i)	280°	=	=	=
(j)	318° 56′	=	=	=
(k)	397°	=	=	=
(l)	450°	=	=	=
(m)	254° 11′	=	=	=
(n)	630°	=	=	=
(o)	371° 3′	=	=	=
(p)	900°	=	=	=

5. Using tables express in degrees and minutes the angles with the following radian measures.

(a)	0.4084 = 23° 24′	(e)	0.9186 =	(i)	0.6013 =		
(b)	1.0996 =	(f)	0.5086 =	(j)	0.9989 =		
(c)	0.0820 =	(g)	0.2272 =	(k)	1.3821 =		
(d)	1.5167 =	(h)	0.4186 =	(l)	1.5432 =		

6. Using tables and the radian and degree angles given in the diagram on page 4, express in degrees and minutes the angles with the following radian measures.

(a)	1.8748	=	1.5708 + 0.3040	=	90° + 17° 25′	=	107° 25′
(b)	2.9817	=		=		=	
(c)	1.6238	=		=		=	
(d)	3.5298	=		=		=	
(e)	3.9907	=		=		=	
(f)	4.6010	=		=		=	
(g)	4.9899	=		=		=	
(h)	5.7626	=		=		=	
(i)	6.0002	=		=		=	
(j)	7.3424	=		=		=	
(k)	8.6961	=		=		=	
(l)	9.9330	=		=		=	

7. Find, in degrees, the third angle of a triangle when two angles are given.

(a) $\frac{\pi}{6}$, $\frac{\pi}{4}$ Ans. ☐ (c) $\frac{5\pi}{18}$, $\frac{\pi}{5}$ Ans. ☐ (e) $\frac{7\pi}{15}$, $\frac{\pi}{18}$ Ans. ☐

(b) $\frac{\pi}{3}$, $\frac{3\pi}{5}$ Ans. ☐ (d) $\frac{4\pi}{9}$, $\frac{5\pi}{12}$ Ans. ☐ (f) $\frac{2\pi}{3}$, $\frac{\pi}{9}$ Ans. ☐

ENVELOPES

Now isong the fact that the Left Hand Edge and th Centre Line of Mathomat are perpendicular to each other, Mathomat can be used to construct the following envelopes.

A The Envelope of a CIRCLE

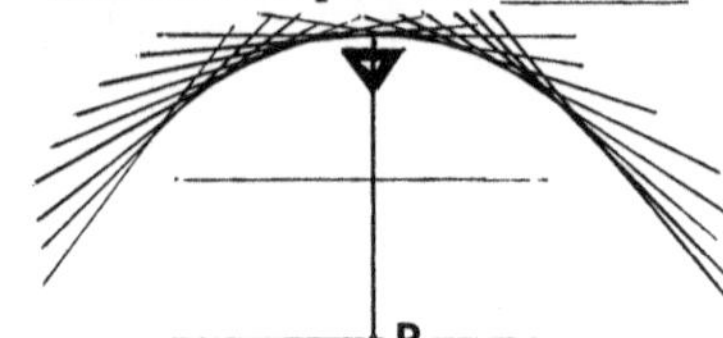

1. Mark Point P.
2. Select any suitable point on the Centre Line of your Mathomat.
3. Keep this point on point P and using the Left Hand Edge draw a series of lines (tangents) around pont P as shown
4. The more lines drawn the closer the envelope will appear as a CIRCLE.
5. Complete the unfinished envelope.

B The Envelope of an ELLIPSE

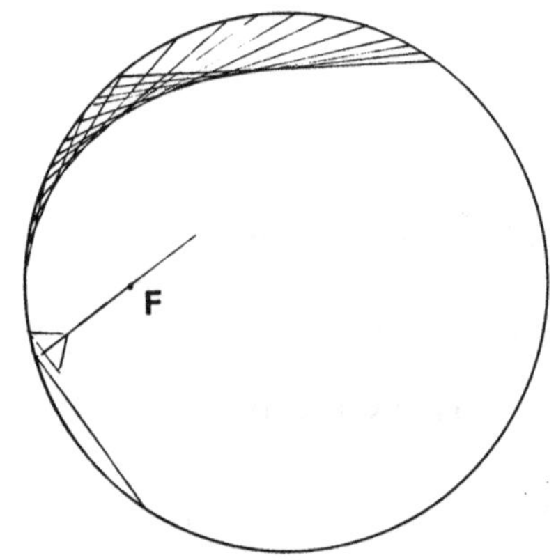

1. Draw any suitable circle.
2. Mark a point F approximately 1cm from the circumference of the circle as shown.
3. Now keeping

   (i) the Centre Line of a Mathomat over pont F
   and (ii) the intersection point of the Centre Line and the Left Hand Edge on the circumference, move around the circle drawing chords as shown.
4. The more chords the closer the envelope will appear as an ELLIPSE.
5. Complete this unfinished envelope.

C The Envelope of a PARABOLA

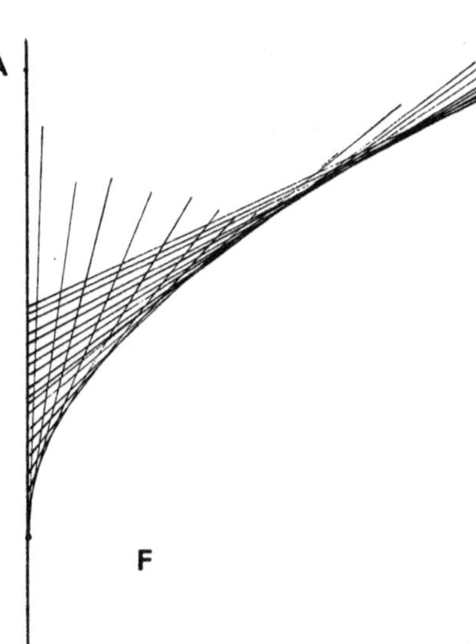

1. Draw the line AB
2. Mark in the approximate centre of AB
3. Approx. 1cm to the right of this point, mark a point F.
4. Now Keeping
   (i) The Centre Line of Mathomat over point F.
   & (ii) the intersection point of the Centre Line and the Left Hand Edge on the line ABmove along AB dawing in suitable lines (tangents).
5. The more lines drawn the closer the envelope will appear as a PARABOLA.
6. Complete this unfinished envelope.

ENVELOPES

D. The Envelope of a HYPERBOLA

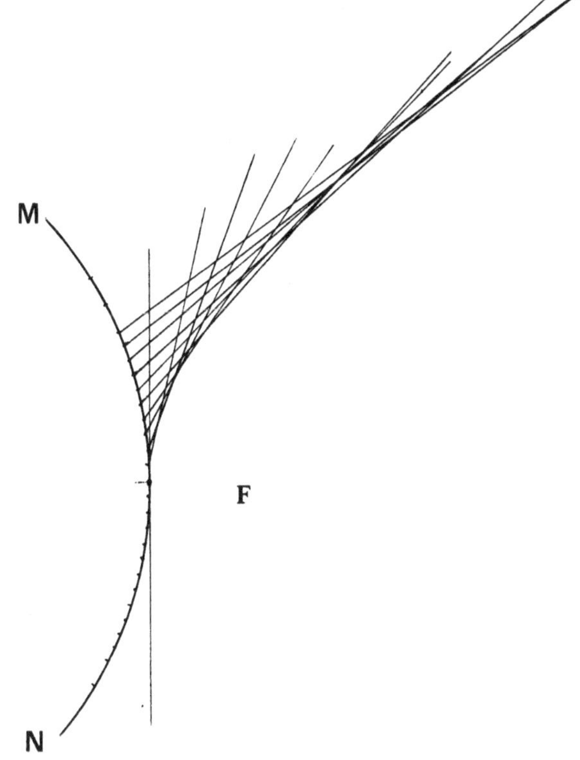

1. Draw the arc MN.

2. Mark in th apporox. centre of arc.

3. Approx. 1 cm to the right of the point, mark a point F.

4. Now keeping
   (i) the centre Line of Mathomat over point F.
   & (ii) the intersection point of the Centre Line and Left Hand Edge on the arc MN, move around the arc drawing in lines (tangents) as shown.

5. The more lines drawn, the closer the envelope will appear as a HYPERBOLA.

6. Complete this unfinished envelope.

The Left Hand Edge of Mathomat may be used as a straight edge to construct the following envelopes.

Finish these incomplete envelopes:

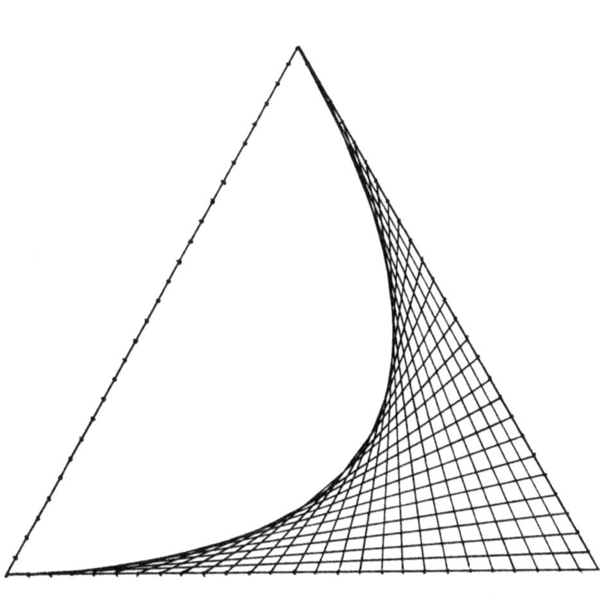

ENVELOPES

Mathomat can readily be used to draw the envelopes of any shape which is bounded by straight lines.

This is achieved by using suitable circles, as shown opposite in the drawing of the envelope of a rectangle.

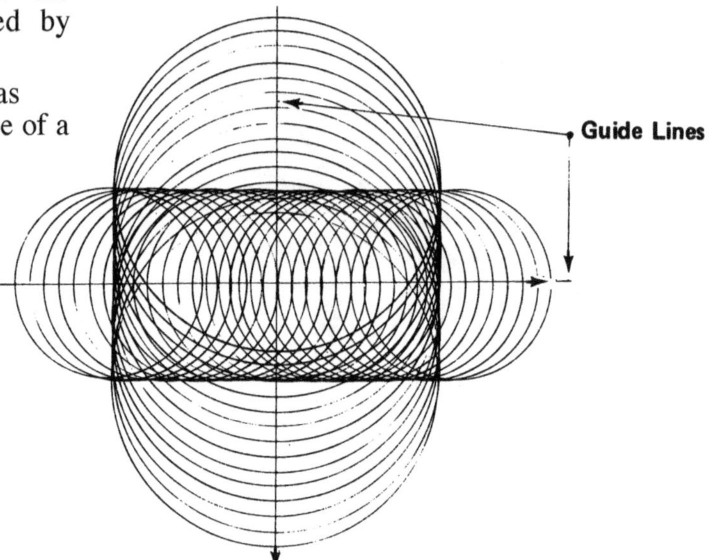

Now discover what shapes are produced when you use the circles in our Mathomat to complete the following envelopes. In each case nae the shape produced.

NOTE: For the best effect, move the circles along the guide lines at regularly spaced intervals.

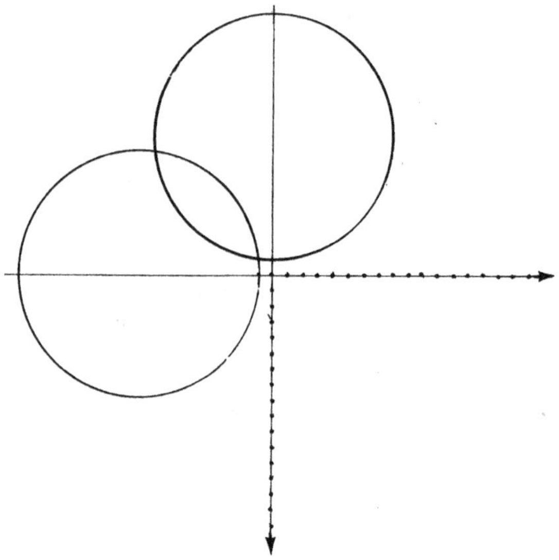

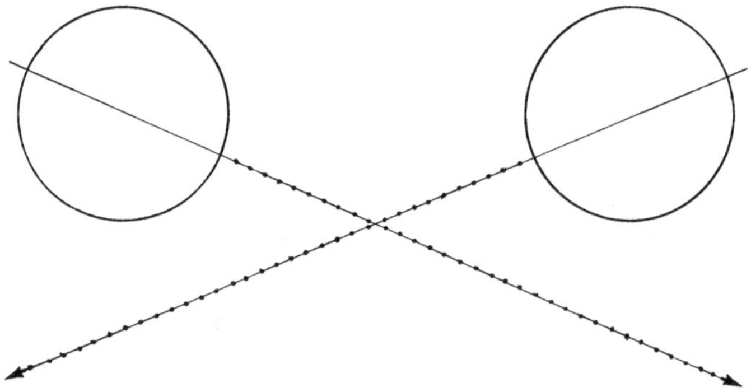

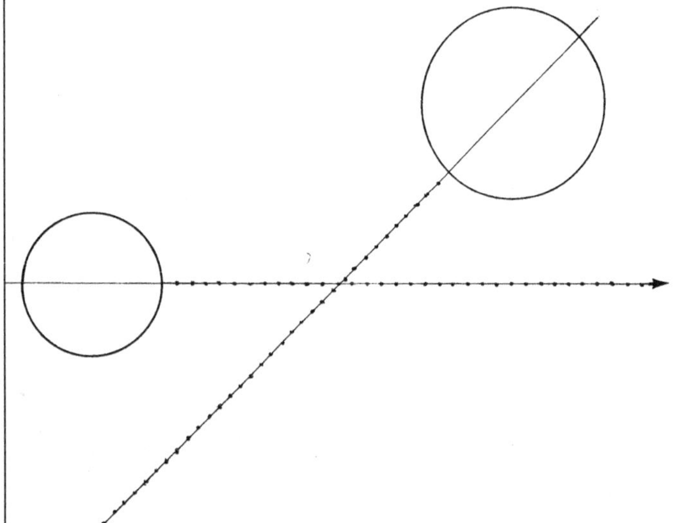

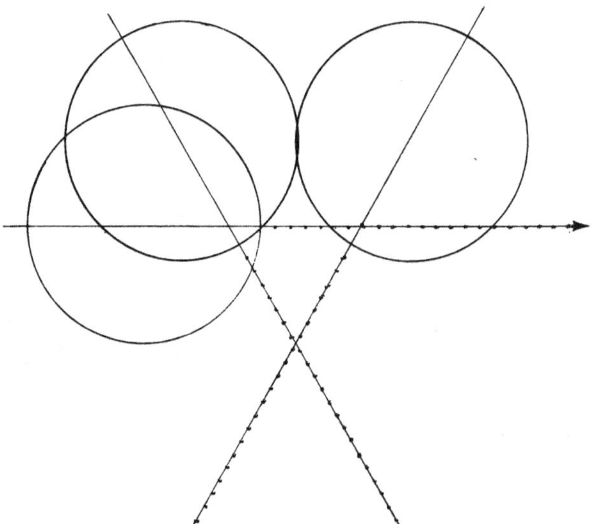

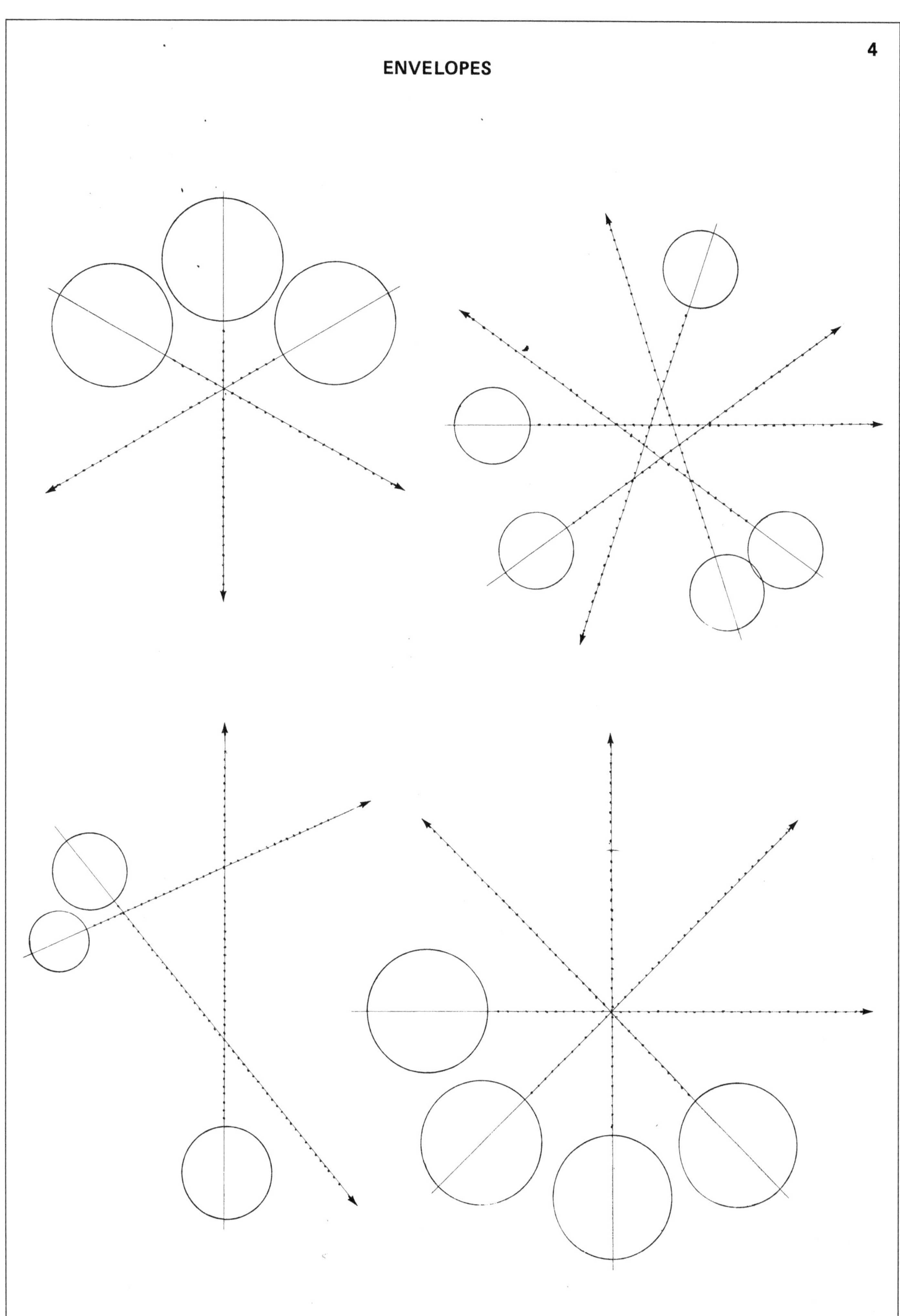
4
ENVELOPES

www.ingramcontent.com/pod-product-compliance
Lightning Source LLC
Chambersburg PA
CBHW061106210326
41597CB00022B/3993

* 9 7 8 0 9 9 4 1 6 1 3 2 1 *